THE MAP THAT CHANGED THE WORLD

ALSO BY SIMON WINCHESTER

*The Man Who Loved China**

*A Crack in the Edge of the World**

The Meaning of Everything

*Krakatoa**

*The Fracture Zone**

*The Professor and the Madman**

The River at the Center of the World

Small World

Hong Kong: Here Be Dragons

Pacific Nightmare

Pacific Rising

*Korea: A Walk through the Land of Miracles**

*Outposts**

Prison Diary: Argentina

Stones of Empire

Their Noble Lordships

American Heartbeat

In Holy Terror

* Available from Harper Perennial

SIMON WINCHESTER

✠

THE MAP THAT CHANGED THE WORLD

William Smith and the Birth of Modern Geology

HARPER PERENNIAL

NEW YORK • LONDON • TORONTO • SYDNEY • NEW DELHI • AUCKLAND

Illustrations by
Soun Vannithone

HARPER PERENNIAL

Copyright and illustration acknowledgments appear on page 331.

A hardcover edition of this book was published in 2001 by HarperCollins Publishers.

P.S.™ is a trademark of HarperCollins Publishers.

First Perennial edition published 2002.

First Harper Perennial edition published 2009.

Designed by Kate Nichols

The Library of Congress has catalogued the hardcover edition as follows:

Winchester, Simon.
The map that changed the world : William Smith and the birth of modern geology / by Simon Winchester; illustrations by Soun Vannithone—1st ed.
p. cm.
Includes index.
ISBN 0-06-019361-1
1. Smith, William, 1769–1839. 2. Geology, Stratigraphic—History. 3. Geologists—Great Britain—Biography. I. Title.

QE22.S6 W55 2001
550'.92—dc21
[B] 2001016603

ISBN 978-0-06-176790-6 (Harper Perennial edition)

24 25 26 27 28 LBC 24 23 22 21 20

FOR HAROLD READING

In days of old, old William Smith,
While making a canal, Sir,
Found out how the strata dipped to the east
With a very gentle fall, Sir.
First New Red Sand and marl a-top
With Lias on its border,
Then the Oolite and the Chalk so white
All stratified in order.
Sing, cockle-shells and oyster-banks,
Sing, thunder-bolts and screw-stones,
To Father Smith we owe our thanks
For the history of a few stones.

Source: Anniversary dinner, A. C. Ramsay, 1854

Contents

Map insert follows page 138.

Illustrations

Chapter-Opening Illustrations

Incorporated in eighteen of the nineteen chapter openings (including those of the prologue and the epilogue) will be found small line drawings of Jurassic ammonites—long-extinct marine animals that were so named because their coiled and chambered shells resembled nothing so much as the horns of the ancient Egyptian ram-god, *Ammon.* Soun Vannithone's drawings of these eighteen specimens are placed in the book in what I believe to be the ammonites' exact chronological sequence. This means that the book's first fossil, *Psiloceras planorbis*, which illustrates the prologue, is the oldest ammonite, and is to be found deepest down in any sequence of Jurassic sediments; by the same token the final fossil, *Pavlovia pallasioides*, comes from a much higher horizon, and is very much younger. Much like the epilogue it illustrates, it was fashioned last. It must be said, though, that anyone who flips rapidly from chapter to chapter in the hope of seeing a speeded-up version of the evolutionary advancement of the ammonite will be disappointed: Ammonites—floating, pulsating, slow-swimming beasts that were hugely abundant in the warm blue Jurassic seas—do not display any conveniently obvious changes in their features—they neither become progressively smaller with time, nor do they become larger; their shells do not become more complex, or less. True, some ammonites with very ridged shells do indeed evolve into smoother-shelled species over the ages, but these same creatures then become rougher and more ridged again as time wears on, managing thereby to confuse and fascinate all who study them. Only studies of ammonites from successive levels will reveal sure evidence of evolutionary change, and such study is too time consuming for the chance observer. Ammonites are, however, uniformly lovely; and they inspired William Smith: two reasons good enough, perhaps, for including them as symbols both of Smith's remarkable prescience and geological time's amazing bounty. However: eighteen ammonites and nineteen chapter openings? There is one additional illustration, of the microscopic cross-section of a typical oolitic limestone, which I have used to mark the heading for chapter 11. Since this chapter is very different in structure from all the others, and since much of its narrative takes place along the outcrop of those exquisitely lovely, honey-colored Jurassic rocks known in England as the Great Oolite and the Inferior Oolite, it seemed appropriate and reasonable to ask the legions of ammonites, on just this one occasion, to step—or swim very slowly—to one side.

Prologue: *Psiloceras planorbis*
Chapter One: *Echioceras raricostatum*
Chapter Two: *Amaltheus margaritatus*
Chapter Three: *Dactylioceras tenuicostatum*
Chapter Four: *Harpoceras falciferum*
Chapter Five: *Hildoceras bifrons*
Chapter Six: *Sonninia sowerbyi*
Chapter Seven: *Stephanoceras humphriesianum*
Chapter Eight: *Parkinsonia parkinsoni*
Chapter Nine: *Zigzagiceras zigzag*
Chapter Ten: *Tulites subcontractus*
Chapter Eleven: Oolitic Limestone

Text Illustrations

A Note on the Map Insert

The brilliance of William Smith's achievement can be amply demonstrated by comparing his great map of 1815 with the one produced today by the British Geological Survey. The similarity of so much of the detail—visible even at a scale where much cannot be seen—is proof absolute of the accuracy and prescience of Smith's work, yet does not admit of the one signal difference between the two productions: that while the survey map is the fruit of the labors of thousands, William Smith's map, drawn a century and a half before, is the result of the dedication and determination of one man who worked for almost twenty years, always entirely alone.

Prologue

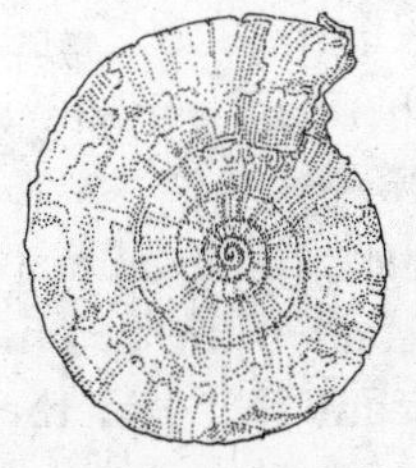

Psiloceras planorbis

Above one of the many grand marble staircases within the east wing of Burlington House, the great Palladian mansion on the north side of London's Piccadilly, hangs a pair of huge sky blue velvet curtains, twisted and tasseled silk ropes beside them. Although many may wonder in passing, rarely does any one of the scores of people who climb and descend the stairs inquire as to what lies behind the drapes. A blocked-off window, perhaps? A painting too grotesque to show? A rare Continental tapestry, faded by the sunlight?

Once in a while someone curious and bold will demand a look, whereupon a functionary will emerge from behind a door marked Private, and with practiced hand will tug gently on the silk ropes. The curtains will slowly part, revealing an enormous and magnificent map of England and Wales, engraved and colored—in sea blue, green, bright yellow, orange, umber—in a beguiling and unfamiliar mixture of lines, patches, and stippled shapes.

"The German Ocean," it says to the east of the English coast, instead of today's "North Sea." There is, in an inset, a small cross-section of what is said to be the underside of the country from Wales to the river Thames. Otherwise all is readily familiar, comfortingly recognizable. The document is exquisitely beautiful—a beauty set off by its great size, more than eight feet by six—and by the fact that it towers—looms, indeed—above those who stand on the staircase to see it. The care and attention to its detail is clear: This is the work of a craftsman, lovingly done, the culmination of years of study, months of careful labor.

At the top right is its description, engraved in copperplate flourishes: "A Delineation of The Strata of England and Wales with part of Scotland; exhibiting the Collieries and Mines; the Marshes and Fen Lands originally Overflowed by the Sea; and the Varieties of Soil according to the Variations in the Sub Strata; illustrated by the Most Descriptive Names." There is a signature: "By W. Smith." There is a date: "Augst 1, 1815."

This, the official will explain, is the first true geological map of anywhere in the world. It is a map that heralded the beginnings of a whole new science. It is a document that laid the groundwork for the making of great fortunes—in oil, in iron, in coal, and in other countries in diamonds, tin, platinum, and silver—that were won by explorers who used such maps. It is a map that laid the foundations of a field of study that culminated in the work of Charles Darwin. It is a map whose making signified the beginnings of an era not yet over, that has been marked ever since by the excitement and astonishment of scientific discoveries that allowed human beings to start at last to stagger out from the fogs of religious dogma, and to come to understand something certain about their own origins—and those of the planet they inhabit. It is a map that had an importance, symbolic and real, for the development of one of the great fundamental fields of study—geology—which, arguably like physics and mathematics, is a field of learning and endeavor that underpins all knowledge, all understanding.

The map is in many ways a classic representation of the ambitions of its day. It was, like so many other grand projects that survive as testament to their times—the *Oxford English Dictionary*, the Grand Triangulation of India, the Manhattan Project, the Concorde, the Human Genome—a project of almost unimaginably vast scope that required great vision, energy, patience, and commitment to complete.

But a signal difference sets the map apart. Each of the other projects, grand in scale, formidable in execution, and unassailable in historical importance, required the labor of thousands. The *OED* needed entire armies of volunteers. To build the Concorde demanded the participation of two entire governments. More men died during the Indian triangulation than in scores of modest wars. The offices at Los Alamos may have housed behind their chain-link fences shadowy figures who would turn out to be Nobel laureates or spies, but they were all hemmed in by immense battalions of physicists. And to attend to all their various needs—be they bomb makers, plane builders, lexicographers, codifiers of chemistry, or measurers of the land—were legions upon legions of minions, runners, amanuenses, and drones.

The incomparably beautiful geological map of 1815, however, required none of these. As vital as it turned out to be for the future of humankind, it stands apart—because it was conceived, imagined, begun, undertaken, and continued and completed against all odds by just one man. All the Herculean labors involved in the mapping of the imagined underside of an entire country were accomplished not by an army or a legion or a committee or a team, but by the single individual who finally put his signature to the completed document—William Smith, then forty-six years old, the orphaned son of the village blacksmith from the unsung hamlet of Churchill, in Oxfordshire.

And yet William Smith, who created this great map in solitary endeavor, and from whose work all manner of benefits—commercial, intellectual, and nationalistic—then flowed, was truly at first a prophet without honor. Smith had little enough going for him: He was of simple yeoman stock, more or less self-taught, stubborn and visionary, highly motivated, and single-minded. Although he had to suffer the most horrendous frustrations during the long making of the map, he never once gave up or even thought of doing so. And yet very soon after the map was made, he became ruined, completely.

He was forced to leave London, where he had drawn and finished the map and which he considered home. All that he owned was confiscated. He was compelled to live as a homeless man for years, utterly without recognition. His life was wretched: His wife went mad—nymphomania being but one of her recorded symptoms—he fell ill, he had few friends, and his work seemed to him to have been without point, without merit.

Ironically and cruelly, part of the reason for his humiliation lies behind another set of faded velvet curtains that hang nearby, on another of Burlington House's many elaborate staircases. There, it turns out, is quite another map, made and published shortly after William Smith's. It was in all essentials a copy, made by rivals, and it was made—if not expressly then at least in part—with the intention of ruining the reputation of this great and unsung pioneer from Oxfordshire: a man who was not gently born, and who was therefore compelled, like so many others in those times, to bear the ungenerous consequences of his class.

✠

But in the very long run William Smith was fortunate. A long while after the map had been published, a kindly and liberal-minded nobleman for whom Smith had been performing tasks on his estate in a small village in Yorkshire, recognized him—knew, somehow, that this was the man who had created the extraordinary and beautiful map about which, it was said, all learned England and all the world of science outside was talking.

This aristocratic figure let people—influential and connected people—know about the man he had discovered. He reported that he was hidden, incognito, in the depths of the English countryside. He supposedly had no expectation that anyone would now ever remember, or would ever recognize, the solitary masterpiece that he once had made. He imagined he was doomed to suffer an undeserved oblivion.

But on this occasion his pessimism was misplaced: The messages that had been sent *did* get through—with the consequence that, eventually, William Smith was persuaded to return to London, to receive at last the honors and rewards that were due him, and to be acknowledged as the founding father of the whole new science of English geology, a science that remains at the core of intellectual endeavor to this day.

It is now exactly two hundred years since William Smith began work on the map that changed the world. What follows, drawn from his diaries and letters, is a portrait of both a long-forgotten man and the world in which he lived and worked, as well as the story of his great map, which has remained hidden behind the blue velvet curtains of a great house in London far too long.

The map that was to become "the map that changed the world." Completed in 1801, this very basic sketch still has an uncanny accuracy.

I

Escape on the Northbound Stage

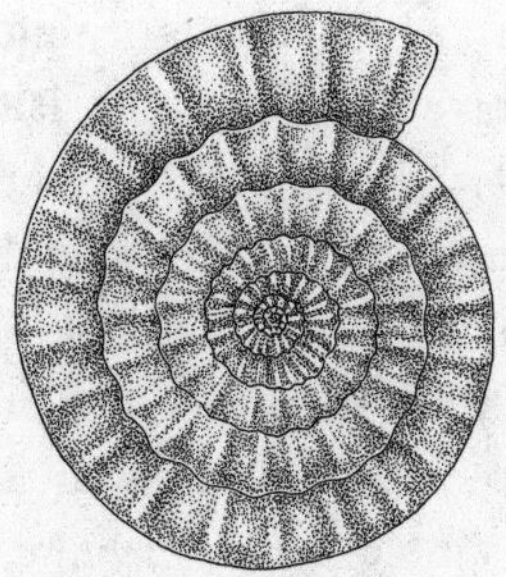

Echioceras raricostatum

The last day of August 1819, a Tuesday, dawned gray, showery, and refreshingly cool in London, promising a welcome end to a weeklong spell of close and muggy weather that seemed to have put all the capital's citizens in a nettlesome, liverish mood.

Anyone trying to hurry along the cobbled and granite-paved streets that day was still certain to be frustrated, despite the improvement in the weather: The crowds! The crush! The dirt! The smell! More than a million people had lately been counted as living within and beyond London's city walls, and each day hundreds more, the morning papers reported, were to be found streaming in from the countryside, bent on joining the new prosperity that all hoped might soon be flowering now that the European wars were over. The city's population was well on the way to doubling itself in less than twenty years. The streets were in consequence filled with a jostling, pullulating, dawdling mass of people. And animals, too: It seemed of little matter to some farmers that there had long been laws to keep them from driving cattle through the center of town—so among the throngs one could spot mangy-looking sheep, more than a few head of cattle, the odd black pig, and of course horses, countless horses, pulling carriages and goods vehicles alike. The stench of their leavings, on a hot week such as this had been, was barely tolerable.

Since it was very early in the morning, there were, of course, fewer crowds than usual. Fewer, that is, except in one or two more notorious spots, where a sad and shabby ritual of the dawn tended to bring out the throngs—and where this story is most appropriately introduced.

✠

The better known of the London sites where the morning masses gathered was in the rabbit warren of lanes that lay near Saint Paul's Cathedral, to the east of where the river Fleet had once run. Halfway along the Fleet Market a passerby would have noted, perhaps with the wry amusement of the metropolitan sophisticate, that crowds had gathered outside a rather noble, high-walled building whose address, according to a written inscription above the tall gateway, was simple: Number Nine.

An onlooker would have been amused because the address was a mere euphemism, the building's real purpose only too well known. The streets to the west of Saint Paul's were one of the two districts of nineteenth-century London where a clutch of the capital's many prisons were concentrated: the Newgate, the Bridewell, the Cold Bath Fields, and the Ludgate jails had all been built nearby, in what in winter were the chill gloom and coal-smoke fogs of the river valley. And Number Nine was the site of the best known of them all, the prince of prisons, the Fleet.

There was another, precisely similar, ghetto of prisons on the south side of the Thames, in the area that, then technically beyond London, was the borough of Southwark: another small huddle of grim, high-walled mansion houses of punishment and

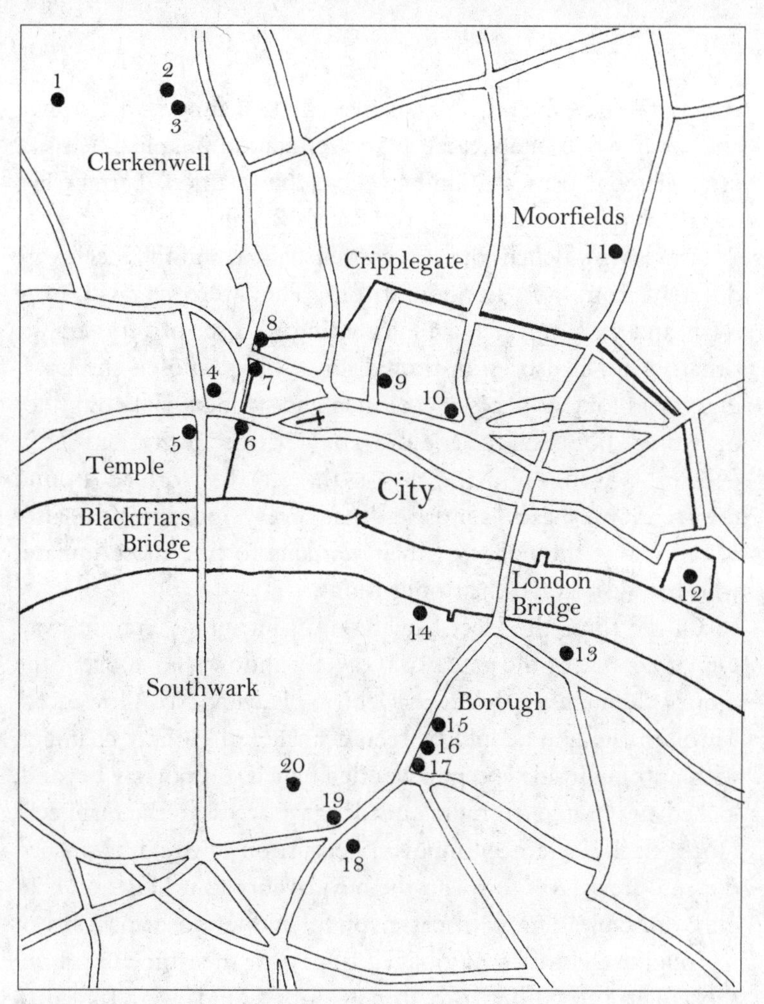

The principal prisons—including those for debtors—in London, 1819. Smith was languishing in Southwark, in what is noted here as King's Bench Prison II.

1. Cold Bath Fields
2. Clerkenwell Bridewell
3. New Prison, Clerkenwell
4. Fleet Prison
5. City Bridewell
6. Ludgate Prison I
7. Newgate Prison
8. Giltspur Street Compter
9. Wood Street Compter
10. Poultry Compter
11. Ludgate Prison II (after 1760)
12. The Tower
13. Borough Compter
14. The Clink
15. Marshalsea Prison
16. King's Bench Prison I
17. White Lion Gaol
18. County Gaol for Surrey
19. King's Bench Prison II (after 1758)
20. St. George's Fields Bridewell

restraint—the Clink, the Marshalsea, the Bedlam prison-hospital, and, formidable in appearance and reputation, just like its sister establishment back at Number Nine, the infamous barrackslike monstrosity of the Prison of the King's Bench.

The King's Bench, the nearby Marshalsea, and the Fleet were different from most London prisons. They were very old, for a start, and were privately run according to a set of very strange rituals. They had been instituted for a sole purpose—the holding, for as long as necessary, of men and women who could not or would not pay their bills. These three institutions were debtors' prisons—and the reason that crowds formed around their entrances each sunrise is that, every morning just after dawn, it was the policy of their wardens to free those inmates who had discharged their obligations.

Of the three the Fleet had the most intriguing entranceway. On either side of the gate was a caged window, and above it the motto "Remember the Poor Debtors, Having No Allowance." Through the grate could be seen a small and gloomy chamber, with nothing inside except a wooden bench. A doorway beyond, locked and barred from the outside, gave access to the main cellblock. Each day a new impoverished prisoner would be pushed out into the cage—to spend the next twenty-four hours on begging duty, pleading with passersby for money to help in his or her plight. Debtors were obliged to pay for their time in prison; those who turned out to be totally out of funds were forced to go into the grated room and beg.

The crowds outside the Fleet and the King's Bench prisons on that cool August Tuesday morning, and that so interrupted the progress of men of affairs on their ways along the granite setts with which the road in Southwark and Saint Paul's had recently been paved, were there to see a spectacle. Tourists came to the jails to see the beggars; the merely curious—as well as the small press of family and friends (and perhaps some still-unsatisfied creditors)—came to greet with amiable good cheer the small group of inmates who each day would emerge, blinking, into the morning sunlight.

✣

According to the prison records, one of the half dozen prisoners who stepped free from behind the high walls of the King's Bench Prison on that Tuesday morning was a sturdy-looking yeoman whose papers showed him to have come from Oxfordshire, sixty miles west of London. Those few portraits painted of him in his later years, together with a single silhouette fashioned when he was in his dotage, and a bust sculpted in marble more than twenty years later, show him to be somewhat thickset, balding, with a weatherbeaten face.

Some less charitable souls might call him a rather plain-looking man, even, in truth, a little ugly. His forehead slants backward, a trifle alarmingly. His nose is somewhat too large for comfort. His mutton-chop whiskers are wayward. But in most of the pictures he seems to be wearing an expression that serves by way of compensation for the facial shortcomings: He seems, from his looks, at once tolerant, kindly, and perhaps even vaguely amused by the droll complexities of his life.

At the time of his release from jail he was fifty years old, and he must have emerged from the main gate into the Southwark crowds that day in an embarrassed and fretful state. He had good reason to be anxious: The previous four years of his life had been trying, racked by debt and uncertainty, by privation and public humiliation. And, as he was soon about to learn, only a matter of hours after his release, his trials were far from over.

The address of his lodgings was given in bankruptcy court as number 15, Buckingham Street, and it was to this imposing stone mansion, in which he had lived for the previous fourteen years, that he now walked, alone. He had spent the better part of the last ten weeks in the miseries behind the bars of the King's Bench, living for most of that time in a crowded cell, a chummage, with two or three others similarly ruined. Now he had his freedom and the pleasure of his own company. He quickened his pace—he was a staggeringly fast walker—as he strode steadily westward to his house.

It was a short enough walk. He had his choice of bridges across the Thames, and had only to turn left when he had made it over the fetid and polluted river with its muddily inelegant banks. He walked steadily along the entire length of the Strand—newly outfitted with the cast-iron lamps of the Gas-Light and Coke Company—and past familiar churches, shops, tailors, and alehouses.

The streets here, by now some distance from the prisons, pulsated with all the elegance and gaiety of Regency times. This, after all, was the day of Beau Brummell (though Brummell himself had only three years before left London for France, preparing for his own date with debtors' prison—and his subsequent death in a French lunatic asylum). The street that morning would have been crowded with the dandies who (their newly invented umbrellas sheltering them from the showers) followed the strict particulars of his style.

The entire stretch along which the glum but relieved Oxfordshire convict walked spoke all too gaudily of money and amusement and *brio*—a sharp contrast, no doubt, to the grim mood of the man who passed among them. In ordinary circumstances he might have stopped at number 181, the elegant bow-windowed building where his best friend, the noted cartographer John Cary, had his offices; but this particular morning he was in a hurry and eager to move on.

It took him ten minutes to pass the length of the Strand, after which he turned off left—Trafalgar Square had not yet been built to act as a landmark—and into that small maze of fashionable Georgian neoclassical houses that had been put up by the four Adam brothers half a century before and that they had named, after *adelphoi*, the Greek word for brothers, "the Adelphi."

Down he strode, past the Savoy, along John Adam Street, and finally into Buckingham Street itself, and up to the front door of number 15.

The door to his house, he was shocked to find, was shut and bolted. A tipstaff stood outside, on sentry duty, and there was a notice pinned to the woodwork: The landlords had repossessed the house and had emptied it of much of its furnishings and papers. Work was still going on; the officer was on hand to ensure that no one—and in particular this one man—attempted to gain entry.

To make doubly certain, the bailiff asked for the man's name. William Smith, the arrival replied. There was an expression of mumbled regret, and the burly sentry took up a stance with his arms folded in front of him, brooking no argument. No, he could not come in. William Smith, beaten down yet again, but now determined not to suffer the indignity of confrontation, turned away.

There was a particularly cruel and desperate irony in this situation. On the following morning, the Wednesday, the same John Cary at whose offices Smith might well have chosen to call, was due to publish in book form the second part of a formidable new collection of geological maps, the latest volume of what was coming to be recognized as one of the most profoundly important books ever made.

Cary's great new *Geological Atlas of England and Wales* had been begun four years before, when the cartographer and his apprentice son, George, had labored mightily to issue a work that was as scientifically epochal as it was physically majestic—the finely engraved, hand-colored map, the eight and a half feet high by six feet wide triumph of cartographic brilliance that was formally called *A Delineation of The Strata of England and Wales with a part of Scotland*, but that has been known ever since as the first large-scale national geological map.

It was a document that was to change the face of a science—

indeed, to create a whole new science—to set in train a series of scientific movements that would lead, eventually, to the inquiries of Charles Darwin, to the birth of evolutionary theory, and to the burgeoning of an entirely new way for human beings to view their world and their universe. The inevitable collision between the new rationally based world of science and the old ecclesiastical, faith-directed world of belief was about to occur—and in the vanguard of the new movement, both symbolically and actually, was the great map, and now this equally enormous atlas that John Cary of the Strand was about to publish, and the revolutionary thinking that lay behind their making.

Both works were the creations of William Smith, the yeoman from Oxfordshire, who, on the very eve of publication, was now being turned away from the marble steps in front of the house in which he had lived and worked for so long. And yet, however much of an embarrassment this must have been, the situation was not wholly unfamiliar to him. For although William Smith's years of careful observation and his wholly innovative ways of thinking were about to alter the course of scientific inquiry forever, he had at the same time been forced to wage what must have seemed a ceaseless war against his own humiliation and ill fortune, forced to waste his energies raging against the cheating and class discrimination that seemed, time and again, to frustrate him.

And here he was now, without a home, without possessions, without any evident future—and yet with his new book, his new great work of science, his masterpiece of craftsmanship and endeavor, about to be offered to the public once again. His situation must have seemed grim indeed, and the brutality of coincidence can hardly have escaped him.

✢

Precisely how William Smith reacted during the rest of that wretched day goes unrecorded. It would be tempting to suppose that he marched swiftly back up to John Cary's office that very afternoon and borrowed money, taking an advance against the sale of his atlas, which would be published the following morning. All that is known, however, is that he decided there and then that he would turn his back on the London that, in his view, had so contributed to his ruin.

So he found and collected his wife, he found and collected the nephew who also then lived with them, and he gathered together what few possessions the two of them, in being turned with evident haste out of the Adelphi house a few days before, had managed to save for him. He made his way across the crowded capital to the Black Swan Inn at Holborn, which was known as the principal stagecoach terminal for travelers making their way to Edinburgh via the Great North Road.

During the summer there was a northbound coach every other day, and if he was lucky* he might even on that same night have won three seats and have been thundering northward in a rocking carriage behind the four great fire-breathing horses of the Northern Mail. His driver would carry him and his fellow passengers maybe sixty miles a day, and so the next morning would see him at Peterborough, then Stamford, then Grantham.

Finally the coach reached the small Yorkshire post town of Northallerton, and this is where, bone weary and hungry, William Smith finally got down and began the process, much like any itinerant tradesman or journeyman, of looking for custom and for work.

"The man might be imprisoned—but his discoveries could not be," he was to write some years later.† "London quitted with disgust. The cheering fields regained."

---

*All too little is known about these particular days in Smith's life, since his diary, normally filled with even the most mundane details of his life, remains blank and abjectly silent. Only circumstantial evidence, together with the writings of his nephew and his own reminiscences written many years later, allows us to hazard a guess at how Smith functioned during this exceptionally trying time.

†Smith made a stuttering attempt at an autobiography very late in life: He made pages of notes, from which these remarks are drawn.

✠

It was to be twelve years before William Smith returned to spend much time in London. The man who was hurtling and banging his way northward on that summer evening stagecoach, was then at the low point of his life—a life that, when recounted in as full a manner as the evidence allows, turns out to have been more honorable, more deservedly honored, and on a world scale much more important than he, at that moment, could have imagined.

2

A Land Awakening from Sleep

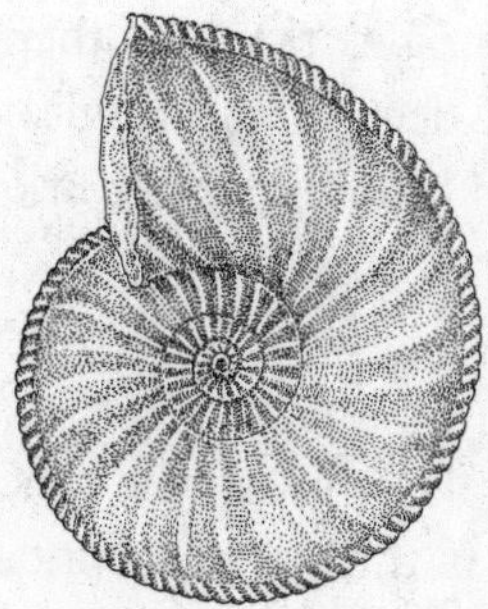

Amaltheus margaritatus

William Smith was born into a world of dogma, faith, and certainty, into a conservative English society that his own discoveries and theories would one day help shake to its very foundations.

And yet already—however conservative the mood of the early eighteenth century may have seemed—there are signs that, viewed from today's perspective, suggest that even at the time of his birth it was imperceptibly readying itself for all that discoverers like Smith would find and do. In countless ways, both great and small, the faiths and certainties of centuries past were being edged aside, and the world was being prepared, if gently and unknowingly, to receive the shocking news of scientific revelation.

Not that any of the vague subtleties of coming change had reached very far. William Smith was born, the first son of the local blacksmith in the hamlet of Churchill in Oxfordshire, on March 23, 1769. It was a measure of the rigor and certitude of both the place and the times that there could be great canonical precision about such a moment. To religious folk—and most English country folk of the time were religious, their daily rhythms set by the steeple bell, their manners regulated from the parsonage—the event that gave a firstborn child to John and Ann Smith in their cottage on the edge of the green in Churchill took place, according to their implacably held beliefs, exactly 5,772 years, four months, and sixteen days after the creation of the world.

Any student of the Bible could have been quite certain about this figure—in fact he or she could have been quite certain as to the very number of *hours* since the Creation, had the Churchill midwife been scrupulous enough to note the time of the infant's birth. A quick calculation could be made on the basis of an almost unchallenged belief about human origins that was then held by most men and women who lived deep in the English shires—the notion that the world had been brought swiftly into existence exactly 4,004 years before the birth of Christ.

✛

Lest anyone forget, all the Bibles that were in use at the time had dates printed in bold scarlet letters in the margins, annotations to the verses of the Old Testament, designed to act as a gentle reminder. "In the beginning, God . . ." had the number "4004 B.C." written beside it; the text of the Holy Scriptures' subsequent dramas, from Cain and Abel onward, had progressively lower and lower red-printed figures in the margin, until the events in the manger in Bethlehem, by which time the figure had been cycled down to zero.

The dating of the Bible was very much an idea of the later Middle Ages. It had taken decades for anyone to come up with credible numbers. In an effort to do so, scores of scholarly zealots had carefully analyzed the basic biblical idea—which had never, after all, volunteered an age for the earth, merely the manner in which it had come about—by sedulously counting the number of human generations they believed to have come and gone between the making of Adam and the begetting of Christ. On the basis of their workings it was reckoned, at the close of the sixteenth century, that the world was, give or take, six thousand years old.

It was left to the genial Irish prelate James Ussher, while he was bishop of Armagh, to fix the date with absolute precision. According to his workings, which he managed to convince his clerical colleagues were impeccably accurate, God had created the world and all its creatures in one swift and uninterrupted process of divine mechanics that began on the dot of the all-too-decent hour of 9 A.M., on a Monday, October 23, 4004 B.C.

The cynical and the skeptical may need some reminding of the fine print—of just what was preached in the church in which William Smith was baptized, of the kind of firm beliefs with which his community was invested. Whatever interest Smith the man might later develop in fossils, geology, and the makings of

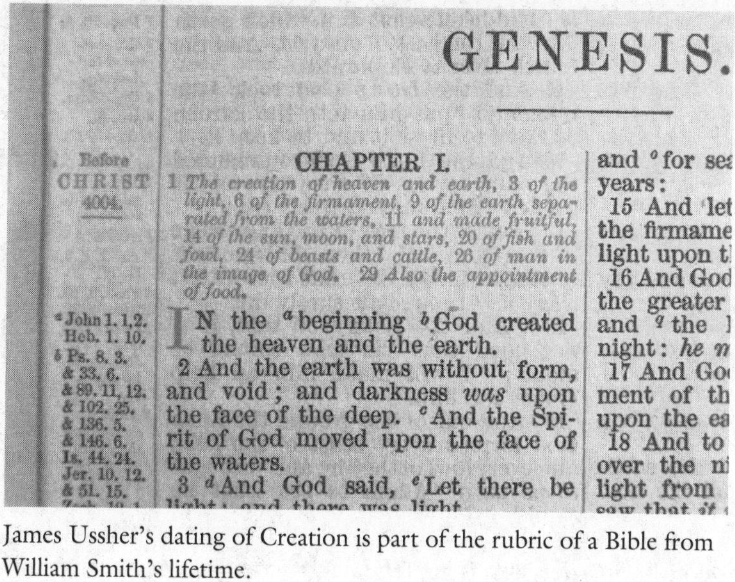
GENESIS.

Before CHRIST 4004.

CHAPTER I.

1 *The creation of heaven and earth,* 3 *of the light,* 6 *of the firmament,* 9 *of the earth separated from the waters,* 11 *and made fruitful,* 14 *of the sun, moon, and stars,* 20 *of fish and fowl,* 24 *of beasts and cattle,* 26 *of man in the image of God.* 29 *Also the appointment of food.*

[a] John 1. 1, 2. Heb. 1. 10.
[b] Ps. 8. 3. & 33. 6. & 89. 11, 12. & 102. 25. & 136. 5. & 146. 6. Is. 44. 24. Jer. 10. 12. & 51. 15.

IN the [a] beginning [b] God created the heaven and the earth.
2 And the earth was without form, and void; and darkness *was* upon the face of the deep. [c] And the Spirit of God moved upon the face of the waters.
3 [d] And God said, [e] Let there be

and [o] for sea
years:
15 And let
the firmame
light upon t
16 And God
the greater
and [q] the
night: *he m*
17 And Go
ment of th
upon the ea
18 And to
over the ni
light from

James Ussher's dating of Creation is part of the rubric of a Bible from William Smith's lifetime.

humankind, at the time he was born there was no question: The entire process of Creation had taken God the familiar six days, and he had begun it 5,772 years before.

At the start of that late October week, in the year that a modern Christian calendar would style 4004 B.C., the Deity organized the basic concepts of light and dark, sun and moon, wet and dry. He then made every ocean, inlet, river, sandbar, meadow, desert, mountain, icecap, and fjord: The structure of the world, its topography, and the geology that forms the core of this story were complete. By the morning of the twenty-sixth, the Thursday, God had seen to it that life had been begun, and by that evening every first microbe, newt, spider, serpent, eagle, cat, horse, and monkey had been duly set in place, to creep, crawl, swim, fly, leap, spring, and deploy its opposable thumb to climb.

By the following day the botanical phyla were all in place: Every rain forest, grassland, savanna, peony, orchid, rose, palm, apple, pine, and daisy had been left on earth, contentedly to bloom. All of Milton's "rocks, caves, lakes, fens, bogs, dens" were now fully accumulated: An earthly paradise was set, ready to be lost.

And by the Saturday, most important of all, emerged those creatures who would lose it. The first two examples of *ur*-human, in the bipedal and upright (but otherwise subtly different from each other) forms of Adam and Eve, had been created in the Garden of Eden. They were at this stage blissfully unaware, of course, and therefore untroubled by the Fall (which would come later, via the agency of the already created serpent and apple).

Recorded history could now formally begin. Human beings were in place, made in the image of their Maker, and they could do with their world more or less as they and their Maker between them pleased. Thus was it all done. Come midnight on the Saturday, with all this frantic labor done, the weary Divinity slept, having declared that all he had created was good, and fully ready to begin the adventuring he had ordained for it for the next six thousand years and more.*

Yet, when William Smith was born, the unquestioning acceptance of a notion such as this was beginning to change. There were vague stirrings of enlightenment from among the nation's chattering classes. Some cynical views—in law, criminally heretical ones—that wafted up from the fashionable salons and drawing rooms of London challenged the very likelihood of Divine Creation. Among them was a new notion, still curious and outrageous to most in the eighteenth century, that Earth might in fact be a very good deal older than the human race that inhabited it, such that humankind and its planet might not in fact have been of near-simultaneous origin.

There was no evidence whatsoever for such views—those who doubted Creation were indulging in little more than inspired hunches. In later years the hunches became more certain, and indeed it would be William Smith's discoveries that would go some long way toward confirming them. But at the time he was born they were very much the idle speculations of a tiny group of sophisticates in London. And the capital was a very long way from northwestern Oxfordshire, both in distance and in temper. The muddy and rutted roads that passed across the ridges of the Chiltern Hills, between Oxford and London, did much to keep at bay any such wild and disagreeable ideas as these.

Where Smith was born, among that small muddle of warm-colored stone cottages, with thatched roofs and climbing roses, the village green and the inn and the duck pond and the old

---

*Few outside the world of the rigid Christian fundamentalists today accept the strict interpretation of James Ussher's arithmetic, which he explained in his monumental work of 1658, *Annalis Veteris et Novi Testamenti.* But nonetheless a 1991 survey showed that fully 100 million Americans still believed that "God created man pretty much in his own image at one time during the last ten thousand years," and anecdotal evidence now suggests that this number is climbing. This might suggest that aspects of the religious climate into which William Smith was born—and that he was to help start changing—are now starting to return.

steepled parish church, beliefs about such weighty matters as humankind's beginnings were unburdened by the complications of too much thought. They were taken on faith as the revelations of Scripture, and when and if they were recounted, they were larded with appropriate and long-remembered quotations from the Book of Genesis.

The infant Smith, whose father and mother were an essentially unremarkable country couple* was thus born into a world of which at least the basis of existence had a certainty. The origins of the planet, just like the origins of mankind, were assumed to be fixed, uncomplicated and divinely directed.

But all such assumptions were to be assaulted, and shockingly so, before the next hundred years were out. To no small degree it was to be William Smith's geological findings, along with a raft of other discoveries, that were to change things. His findings were to prove vitally important in triggering the collision that was eventually to take place between the religious beliefs that were in the ascendant at the time and the scientific reasoning that would provide the spur for the intellectual activities of a century later.

Science was the key—along with the scientific method, with all its underpinnings of observation, deduction, and rational thought. The consequence, once the theories of Charles Darwin in particular had begun to sink in, was a profound modification of the way in which people thought of nature, of society, and of themselves. Which makes it all the more appropriate, given the impact his ideas would have, that it was into a time of suddenly accelerating scientific achievement and technological application that William Smith was born.

For, at the very moment that he was born, things were chang-

---

*Smith was to feel somewhat embarrassed in later years about his forebears' determined ordinariness, and he tried long and hard to prove that through his mother he was a descendant of Sir Walter Raleigh. He convinced no one and eventually abandoned the quest.

ing, and changing fast. In the year of his birth—which according to parish records at Churchill was 1769—there were, for example, three developments, nicely coincident, that in retrospect suggest all too powerfully that change was in the wind. As indeed it was: For the first time in British history the word *industry* was no longer being used simply to describe the nobility of human labor and had come instead to mean what it does today: the systematic and organized use of that labor, generally with the assistance of mechanical devices and machines, to create what would thenceforth be called *manufactured goods.* The Industrial Revolution, in short, was at hand, and three creations from Smith's birth year are well worth noting, since they more than anything suggest the temper of the times. As it happened, for instance, 1769 was the year of grant of patent for James Watt's first condensing steam engine—perhaps the most important invention of the entire era. Josiah Wedgwood, who had been busily making fine pottery in Staffordshire for some years past, opened his great factory, known as Etruria, near Hanley, also in 1769. And the great field of textile making, which was being steadily revolutionized by a cannonade of new inventions, was most notably advanced by the creations of Richard Arkwright—who made the first water-powered cotton-spinning frame, also in 1769.* Watt, Wedgwood, and Arkwright—a holy trinity from the brave new world that was coming into being—were now unknowingly ushering in the man who would change the view of that world for all time.

In all corners of the industrial world there was change, development, innovation, the shock of the new. Coal, iron, ships, pottery, cloth, steam—these were the mantras of the moment. The great English ironmasters, for example, were approaching their zenith: Cranage, Smeaton, and Cort were developing the processes for "puddling" iron and rolling molten metal.

---

*James Hargreaves, whose mechanical spinning jenny was destroyed by fearful proto-Luddites, and Samuel Crompton, whose spinning mule was a hybrid of its two predecessors, came only a little later.

Abraham Darby and John Wilkinson were constructing the first iron bridges in the world. Wilkinson, unarguably the greatest of all eighteenth-century champions of things ferrous, was making the first mine railway in 1767, then the first iron chapel (for a congregation of Wesleyans), and was using iron lighters to shift coal to his three furnaces (and, to cap it all, had himself buried in 1808 in an iron coffin).

Iron production was on the way to doubling every twenty years when Smith was born, and coal was too; and—in what would prove of the utmost significance to William Smith by the time he was a grown man—the mania for canal building, to provide a means of transporting all the coal and iron and finished goods, was teetering at its beginnings.

If there were hints of a coming change in the long-held systems of belief; if the industrial world was accelerating out of all imagination; then so also, and as an obvious corollary, social change was underway as well. And when William Smith was born, the rate and scale of alteration to society was such that even those in so small and isolated a settlement as Churchill, Oxfordshire, would be bound to notice.

Parliament, for example, was in the last decades of the eighteenth century passing enclosure acts at the rate of one a week. The formerly common-held land was now gradually being fenced and hedged, and farmed in a way—with the use of new machines and according to the principles of crop rotation—that led to the creation of the English countryside that we still see today, mannered, orderly, and inordinately pretty.

The village of Churchill itself was still unenclosed in 1769. The local farmers worked the fields as most of England had for centuries, taking for themselves alternating strips of the common-held land and on each strip growing crops, or setting each to pasture, or leaving each fallow, as individual mood and season suggested. The method was woefully inefficient, the landscape it created plain and uninteresting.

But then in 1787, under the usual pressure from the local squirearchy and the more powerful farmers, an enclosure act was passed for both the village and its surrounding countryside. Gone, within a year, were the ragged strips of new-plowed land and the mean acres of wood. The gently dipping fields and meadows that are still to be seen today were all hedged and ditched and ha-ha'd into existence when Smith was still a youngster. It was a development that had profound importance for the English farmer and the English countryside. It was also to be of profound importance for the beginning of career and inspiration for the young William Smith.

There was more to the farming revolution than the fashioning of a handsome landscape. To add luster to the newly made meadows there came new breeds of cattle and sheep—Hereford cows, Southdown sheep among them—that started to be introduced in the late eighteenth century, with the animals at last approximating in appearance (fatter, sturdier, and healthier than their bony and goatlike forebears) the look of the breeds to be seen today. Well-to-do farmers were so proud of their new beasts that they had paintings of them commissioned, and by doing so founded an entirely new artistic school of domestic animal portraiture.

Farming methods improved at a staggering rate, and in consequence the output of grain and potatoes and meat rose hugely. White bread became a commonplace in the diet of rich and poor. Cheese became hugely popular. An abundance of cattle feed all year round meant that at long last the winter ritual of eating only salted beef—the cattle hitherto had all died in the first cold snap for want of feed—could now be ended: A joint of roast beef promptly became a central feature of the national dinner table, part of England's national mystique (and, of course, the Englishman's French nickname, *Le Rosbif*).

And this all led to something else. In fact it was during the late eighteenth century—most probably for the first time—that society suddenly seemed to realize it had become a vastly complicated entity, its characteristics linking and interconnecting with one another in wholly unexpected ways. Such domino effects first became apparent when it was revealed, at the turn of the century, that Britain could no longer feed itself.

The consumption of white bread and roast beef, for example, led indirectly to a set of completely unanticipated consequences. Although the nation's farmers certainly produced a lot—being armed with such weapons as the crop-sowing inventions of Jethro Tull, and the revolutionary land management methods of Thomas Coke, all the benefits of enclosure—and although what they produced, like the bread and the meat, was a delight to eat, it became an unfortunate reality that from that moment on until today, they could not produce enough. England became during this period and for the first time a net importer of wheat and corn.

This was due to the simplest of Malthusian reasons—the fact that the country's population had begun to rise significantly since midcentury. But figures had begun to inch up not because of an increase in birthrate going hand in hand with the rising prosperity, but mainly because of a small but important fall in the nation's death rate. And that was due, in no small part, to the better diet of white bread and roast beef. An unexpected interplay of factors, indeed—all part of the making of Britain as a modern, complicated society, a society readying itself for modern, complicated ideas.

There were other factors in play as well. Health was improving, for example. A child like young William Smith could be more assured than ever before of survival: There was better midwifery, a relative abundance of doctors, the construction of lying-in hospitals for women in labor, the introduction after 1760 of smallpox inoculations, the widespread opening of dispensaries, and a general agreement that fresh air was good for one and that hygiene and ventilation should be regulated—all such developments, all occurring in the latter half of the eighteenth century, helped to ensure that childbirth was far less risky an adventure than before.

Moreover, people simply knew much more than before. Their lives were more efficient and comfortable than they had ever been. There was ample reason for a new degree of physical contentment—an atmosphere that, for those who were so predisposed, was highly conducive to study, to pondering and wondering. There had been steady improvements in education and literacy (Samuel Johnson's great *Dictionary* had been published in 1755). There was now a mature newspaper industry. The postal system was becoming reliable and even efficient—a letter mailed in London could reach Chipping Norton, which was close to Churchill, the afternoon of the following day, "on every day except Monday"—meaning that people, even in so remote a part of the country as Oxfordshire, could now keep abreast of national developments, could tap into an ever-running wellspring of advice and information.

They could learn, and by comparison with what had gone before, they could learn in double-quick time, something of the trivia of trends—as when eighteenth-century gentlemen farmers were beginning to buy pianos for their newly carpeted living rooms. They could know how a Mr. Chippendale began to turn out enchanting new styles of furniture from a new wood, mahogany, which had been discovered in South America. They could read how ladies in Liverpool, Manchester, and Edinburgh were starting to supplement their inelegant skirt pockets by carrying with them what they would call "indispensables," which would be later called handbags. People in Churchill knew that young ladies of fashion, reading the new colored style journals, were now preferring to sport interestingly pale faces instead of the sunburned cheeks of the peasantry. The women of Churchill could learn all too rapidly how—in part to achieve this look—the recently invented parasols and umbrellas were becoming "quite the thing."

And they could learn of foreign developments—the rising agitation in the Americas being the most vexing—or of the minutiae of their own national government (George III, the capricious and unstable farmer-king who had assumed the throne in 1760, oversaw no fewer than seven governments during just the first decade of his reign).* The population now could and did display its anger and its pleasure at matters of which it came to know. The people could rant against unfairnesses—the naval press-gang, say, which was still much in operation in the port cities. They could cheer and argue over the spread of civil rights—John Wilkes, the "Friend of Liberty," was a prisoner in the Tower† when Smith was born; Thomas Paine was marshaling the ideas that would eventually lead him to write *The Age of Reason*; Edmund Burke was well into his career as the foremost liberal thinker of his time.‡

By 1781—by which time William Smith was a twelve-year-old boy—Samuel Johnson was calling the English "a nation of readers." Few were the major towns that did not have a library. Few were the shop signs in the streets that did not show the name of the merchant instead of merely a picture of what he sold. It was assumed, and with reason, that sufficient numbers of passersby would have no difficulty reading the words on the boards—something that preceding generations (and many on the Continent even then) would have found a considerable challenge.

---

*William Smith was born during the administration of the sixth and least distinguished, the duke of Grafton, who acted as caretaker between the administrations of William Pitt the Elder and Lord North.

†The radical politician in whose memory the famous actor Junius Brutus Booth named the son who would assassinate Abraham Lincoln in April 1865.

‡There is a small, Smith-related coincidence here. Edmund Burke made what was perhaps his most famous speech in 1788 when he was opening for the Commons the impeachment proceedings against Warren Hastings, the governor-general of India who, by a coincidence of which the Smith family was only too well aware, had also been born in Churchill. Unlike Smith's small cottage on Junction Road, the house in which Warren Hastings was born still stands. There is some greater fairness in the nomenclature of contemporary geography, however—notably the existence in modern Churchill both of a Hastings Hill and a William Smith Close.

No matter the outcry that allowing the working classes to become educated was to debauch them and tempt them to abandon the manual labors for which they were best suited. "Nineteen in twenty of the species were designed by nature for trade and manufacture," said a writer in *The Grub-Street Journal* at the time of Smith's birth. "To take them off to read books is the way to do them harm, to make them not wiser or better, but impertinent, troublesome and factious." That kind of thinking was rapidly to become outmoded during the years when Smith was growing up: Whatever the political outcome—whatever the effect of the new phenomenon of public opinion, which literacy, communication, newspapers, and libraries encouraged—the nation, save for its most reactionary elements, seemed generally prepared to come to terms with the new mood for change.

✢

William Smith's formative years unrolled through a period that was both astonishingly vibrant and deeply challenging. Advances were firmly under way in almost all applied areas of science and philosophy, and in social change and artistic endeavor as well. But there was still a terrible hesitation about humans' understanding of the most fundamental questions of why they were where they were, who had placed them there, what was the point, what were their origins, what was their fate?

The hesitation was deep rooted; it stemmed, at least in part, from the frank reluctance of eighteenth-century men and women to accept that there even *was* a need to know and wonder at such things. To inquire with true rigor into matters that lay at the heart and soul of his and all society's beliefs smacked, indisputably, of heresy. Even by the time that young William Smith was starting to take advantage of the world's new and inquiring mood, there was still the wide acceptance—not yet contradicted by any evidence that seemed to matter—that God had created both human beings and all the world in which they lived. That was that: No more needed to be said.

And yet. A very few bold and more radically inclined thinkers—Joseph Priestley, one of the discoverers of oxygen, and Erasmus Darwin, Charles's grandfather, among them*—were beginning, in these same extraordinary years, to take a more muscular and skeptical approach to the received wisdom of the Church. By the time Smith was coming to his maturity, questions about these fundamentals were being asked by more than the mere metropolitan sophisticates. The hunch that God might not have done precisely as Bishop Ussher had suggested, or during the time he calculated, was beginning to be tested by real thinkers, by rationalists, by radically inclined scientists who were bold enough to challenge both the dogma and the law, the clerics and the courts.

There was in those early days much more questioning than there was answering. It was a period more marked by bewilderment than certainty. While most still believed that the Scriptures could comfortably provide answers to all the questions about earthly origin and human purpose, there was a growing and more frequently admitted sense of puzzlement as well—a puzzlement that seems to have been most keenly felt among those scientists and engineers who were observing the natural laws of physics and chemistry, who were working with steam or fashioning iron or digging cuts through cliffs. Among those and others who knew something of the newly formulated laws of science, there was a new mood of questioning that hinted that maybe, just maybe, the old beliefs, rooted in the blind acceptance of churchly teachings, might not have been wholly true.

A febrile fluttering of questioning began—about what exactly

---

*Joseph Priestley and Erasmus Darwin, along with Josiah Wedgwood and James Watt, were all Lunaticks, members of Birmingham's Lunar Society, which met monthly on the occasion of the full moon. Freethinking, radical ideas were welcomed by a group that was principally involved in applying scientific discovery to the newly flourishing world of industry.

was the world? How had it, and all that was in it, really come about? Was it sacrilege to wonder such a thing? Was it blasphemy to ask? Would lightning strike down anyone who questioned the likelihood of James Ussher's numbers being correct? Would plague and boils tear at the vitals of anyone who asked out loud just what story might it be that lay buried in the stones beneath our feet?

And all this questioning tended to coalesce around one new and barely structured field of study and fascination. Could it perhaps be that *geology*,* the frail and stripling science that had first been established to inquire into the nature of the earth before and after the Deluge, could it be that *geological* inquiry might hold the answer? This was a science that, after all, had at least the potential—if it could be divorced from churchly dogma—to at least define and then ask the questions to which answers now seemed so urgently needed.

At the time of Smith's birth, geology and those few men who called themselves geologists saw it as no part of their duties to inquire more fully, to delve more deeply, into what were still seen as the realms of the Divine. And yet some scientists were beginning to wonder if geology really was to be confined like this—if it was obliged to function only within the framework of faith, and not to challenge it one whit—then was it truly worthy of being called a science at all?

Maybe, though, it could rehabilitate itself. Maybe geology was the one new scientific discipline that, if applied courageously, might be able to help answer the fundamental and unasked questions that were beginning to trouble those tentative, nervous questioners. Perhaps geology could be the key for those who, in

*The word is first used in English in its modern sense in 1735, though only rarely—and probably not until 1795 can it be considered a mature and full-fledged concept. There was no mention of geology in the 1797 Third Edition of the *Encyclopædia Britannica*; but the Fourth, which came out in 1810, had a lengthy entry, the science by now fully established.

the enlightened, wondering spirit of the times, were at last beginning to tap their fingertips on the stout door of received belief?

✣

Many Europeans who found themselves in England in the closing decades of the eighteenth century talked of seeing a country "waking itself from sleep." Many in England agreed and wondered out loud: Could it be that in shaking and worrying and waking from its sleep the very land itself, by asking at last what exactly *was* that land, and how it had first come into being—could it be that by doing this they might answer questions that would help lay bare the very core of knowledge?

That was what a few men were at the time beginning to wonder. In turn the wonderment of some of them—a country surveyor here, an Oxford-educated priest there, a fossil-collecting dilettante in this city, a radical-minded landlord in that—would be passed down to the intelligent and inquiring young Oxfordshire lad, who would before long help lay the foundations for a brand-new science that would inquire, quite fearlessly and, eventually, scandalously into the foundations of just about everything. William Smith appeared on the stage at a profoundly interesting moment: He was about to make it even more so.

3

The Mystery of the Chedworth Bun

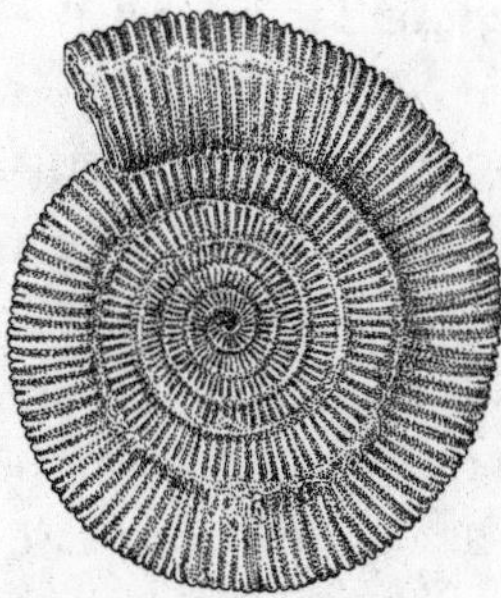

Dactylioceras tenuicostatum

William Smith's introduction to the curious magic of geology—at a time when it was still a calling more nestled in myth and mystery than in the rigors of scientific discipline—came about in the most prosaic of ways.

It was presaged by domestic tragedy. When William was just eight years old his father, John, suddenly died. He had been something of a skilled mechanic as well as the village blacksmith, and had caught a chill while working outdoors repairing a piece of farm machinery. It was a devastating blow, financially as well as emotionally.

For the next couple of years, until Ann Smith married again, the boy, along with his two brothers, Daniel and John, and a sister, Elizabeth,* was largely brought up by his uncle (who, confusingly, was called William too), who also farmed locally.

---

*Elizabeth's own son, as we shall see, was to become professor of geology at Oxford University, due almost wholly to the tutorial inspiration of his uncle William.

A romantic might well say that William Smith was almost an orphan. To all intents and purposes, with a father dead and a mother apparently more interested in her new husband than in her child, he was. Life on his uncle's farm, however, does not appear to have struck him as displeasing. And in any case the farm was itself soon to become symbolically very important in his story—principally because it was substantial enough an establishment to include a dairy.

It was customary in the farms of this part of Oxfordshire for the dairymaid who operated the butter scales (in this case, William's aunt) to use as a pound weight not an artificial metallic object, as one might buy in a market, but one of any number of curious, rather attractive, sometimes flattened and usually almost circular stones that could easily be found in the quarries nearby.

The farmers took these stones for granted—and why not?—they were merely stones of a certain size, one of those very few conveniences of isolated rural life, to be given no more heed than one might give a clod of earth or a muddy pool. But William, who seems to have had a more curious eye than most, saw on the dairy scales objects that were not ordinary at all. He looked carefully at each one, and realized that there were features about these stones that were uncommonly lovely—and had a meaning that no one else seems to have noticed. No one in the dairy, at the very least.

Some features of the pound stones were quite plain to see. Others required a closer look. When viewed from above, the stones appeared to be round, although in fact some of them were not—some had five sides, but sides that were sufficiently chamfered to give the stone, on cursory inspection, a circular appearance. Then again all the pound stones, whether they were actually round or five-sided, were about four inches across. In cross-section the rounded ones were slightly flattened at the top and bottom, so that each stone would sit on the weighing scale without any inclination to roll off. The dairymaids liked this feature: a weighing stone that would stay where it was put.

But whether the stones were flat or indented, round or pentagonal, the Oxfordshire dairymaids found them useful also because they were remarkably uniform in both dimension and mass. Almost all of them weighed in at about twenty-two ounces—which just happened to be what the local dairy managers, in measuring out freshly churned butter for their customers, called a "long pound."

The five-sided variety was arranged in one of two general kinds of shape, each of which can best be imagined by thinking of an orange, either pressed down flat on a table, or else pressed down on top of a small pebble. Both, in other words, were rounded at the top, but some were totally flat below, while others had a shallow upward indentation at the base.

Oxfordshire pound stones, which the locals still sometimes call Chedworth Buns, are still be found in the plowed fields around Churchill. Close inspection of a good specimen reveals them to be even more complicated than one might suppose. The possession of five sides already suggests some degree of complexity; but in addition their outer surfaces are decorated with a series of quite beautiful filigrees of fine lines and beguiling, elegantly regular patterns.

On the top of each rounded dome, for example, is a small disk, composed of five leaflike plates, which surrounds a tiny circular hole. Down what anyone fondling the stones will surely regard as their rather voluptuously convex sides spread ten raylike arrangements of what appear to be plates, like armor. The plates are two sizes, large and small—and they are arranged so that one array of the smaller plates alternates with another of the wider plates, five times each. Underneath, at the point where these rays all come together again, like lines of longitude on a globe, there is another hole, quite larger than the one on top. And on some of the five-sided pound stones—but not on the totally round ones—there is a third hole, somewhat elongated and lozenge shaped, which lies halfway down the curved exterior, right in the center of one of the wider, large-plate rays.

✣

Such things had actually first been recognized for what they were—or what they appeared to be—about a century before. It was in Sussex that a local naturalist, wondering about the version of Oxfordshire pound stones that he found in the fields near Brighton, realized that he had seen something very similar-looking that was actually *living* in the rock pools of his local seashore. Wedged into recesses of the pools he had found scores of almost spherical animals, some of them round, some heart-shaped, all covered with sharp spikes—sea urchins, he knew they were called.

Their protective covering of spikes disguised the exoskeletons beneath. But once in a while the naturalist would find on the beach a dead sea urchin, a specimen that had completely lost its spikes. Such a specimen, naked among the pebbles, displayed its eggshell-white exoskeleton perfectly. And this was the vital point of contact—for although the skeletons that were to be found on a Sussex beach were hollow, and fragile, and soap-bubble light, in shape and size and markings they were precisely the same as the solid, heavy stones he had found in the fields nearby.

So this amateur scientist promptly deduced what we today would regard as perfectly obvious—that his pound stones were stones, yes, in that they were solid and made of what were evidently mineral materials, but they were also clearly the remains, or at least precise simulacra, of common sea urchins. They were members of the family known as echinoderms, and of the genus known as *Echinus*, both of which had been named after the Greek and Latin words from which we get *urchin*, which means "hedgehog."

✠

The symmetry and beauty of the sea-urchin-shaped stone that lay on the dairymaid's butter scales evidently caught the young William Smith's imagination. It is not difficult to imagine him picking it up, turning it over and over in his hand, examining through a glass its intricate patterns of plates, striations, and whorls. He may well have compared it with pictures of modern sea urchins that he found in his textbooks in the village school, and asked questions of his teacher—an eccentric villager named Billy Watts, who seems to have taught his classes while sitting with a cat on each knee, and who was probably not the source of much enlightenment on the subject.

And, most crucially of all, he may have asked himself questions that more disciplined scientists were even then beginning to puzzle over: Just what was a creature of the oceans doing, preserved—*so strangely!*—as part of a rock? Just how did one solid become so firmly embedded inside another? Just what did such things, such weird phenomena, the encapsulation of objects from the sea deep inside the rocks of Oxfordshire, really mean?

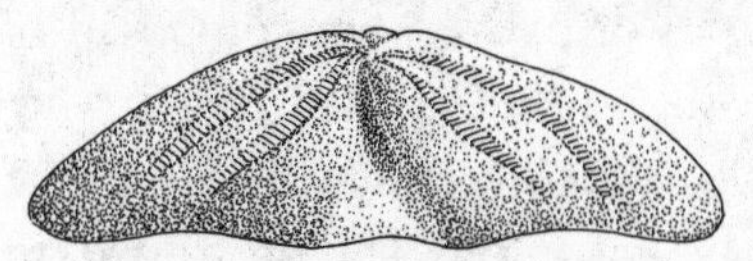

The pound stone, viewed from the side.

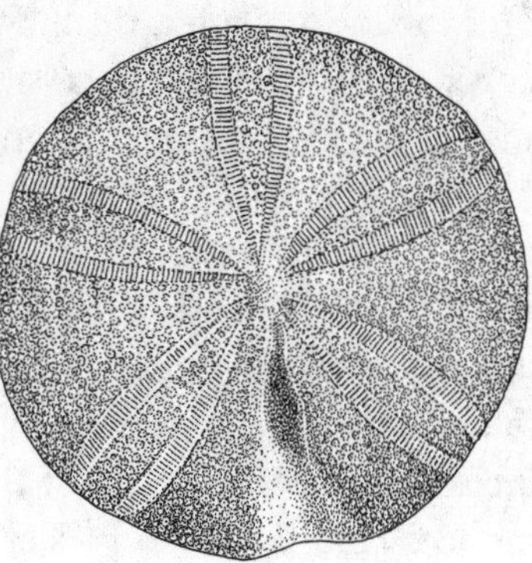

Clypeus ploti—a pound stone—viewed from the top.

It was not long before another such object, just as prosaic, just as lovely, attracted his interest.

Scattered all over the fields around Churchill—but of no interest to the dairymaids—were hundreds of small, thumbnail-size objects, almost spherical, and with a vaguely opalescent sheen to them. They, too, had a subtle, magical beauty: On closer examination, when it was possible to see that their skins had an orange-peel-like texture, with thousands of tiny holes regularly arranged over the surface, their loveliness was even more apparent. They looked a little like acorns; or perhaps William, who might have been taken on school trips to some of the many Roman ruins with which Oxfordshire was littered, might have thought they looked a little like Roman oil lamps. The local farmers, who also spotted the similarity of shape, had long called them lamp shells. William called them "pundibs," and in his diaries written years later recalled using them as marbles, to play games with his schoolmates.

But like the sea urchins, these lamp-shell stones clearly had no logical place among the arid amassments of rock lying beneath the Oxfordshire fields. For these mimic creatures—if that is what they were—likewise belonged to the ocean. Contemporary, living versions, which could also be found on shallow-sloping sands beside the sea and in the estuaries of rivers, were composed of two small shells, the upper one slightly larger than and curving over the lower, with a small hole at the overlapping edge of the upper shell from which a small gelatinous leg protruded to anchor the shell to the ground.

Once in a while, when shallow sea water washed over the tiny animal, a most extraordinary thing happened: The two shells, which were evidently hinged close to the anchor leg, opened slightly, and from between them flicked a long, curled, rubbery tonguelike organ, which waved among the suspended particles in the water, collected some of the edible morsels that stuck to its surfaces, and was then coiled back smartly between the shells, which promptly snapped shut. Following this lightning-quick feast the animal then remained static, evidently digesting and nourishing itself on whatever its remarkable feeding limb had managed to collect.

Today we know only too well exactly what the two objects were that the boy so admired, and that so inspired him. The pound stone, an echinoid, was in all probability a species named *Clypeus ploti*, the genus name (from the Latin) given because of its round, shieldlike shape, and its species name given in honor of a long-forgotten Oxfordshire geologist called Robert Plot.* The marbles that William used in schoolyard play were a type of brachiopod, a terebratulid, and most probably, given its roundness and suitability for games, a small and pretty variety known as *Lobothyris*.

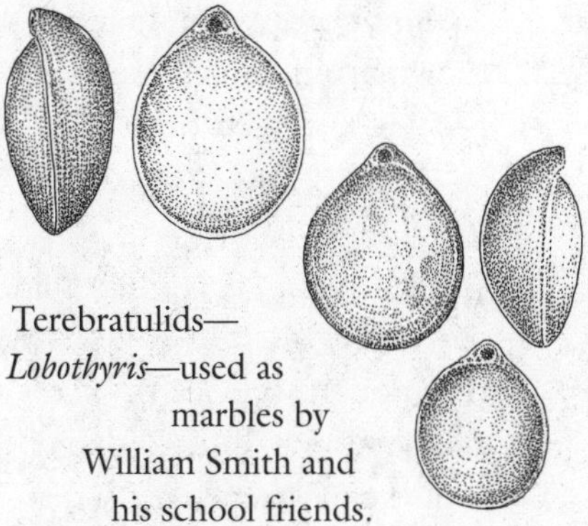

Terebratulids—*Lobothyris*—used as marbles by William Smith and his school friends.

But all this begs the original question: What were such sea creatures doing in the middle of a stretch of unenclosed pasture, a hundred miles from the nearest shoreline and (considering the height of the surrounding land above sea level) a good three hundred vertical feet above it? To answer such questions today is quite simple: The shells are just fossils, once organic but now mineralized relics of a time when the rocks that lie beneath the fields of Oxfordshire were themselves being created, thousands of feet beneath a life-rich tropical sea. But in the late eighteenth century, no such theory had ever been even vaguely imagined. When William Smith was being entranced and captivated by the dairymaids' pound stones, noth-

*Plot is seen by some historians of the science as having been a much-overlooked contributor to our knowledge of the Jurassic era. Naming an echinoid after him, while to outsiders seeming to damn with faint praise, perhaps helps somewhat to redress the balance.

ing about fossils was simple, nothing was universally accepted, nothing was obvious at all.

The entire notion of fossils, in fact—what they were, why they were where they were, what possible deeper meaning was signified by their existence—was quite profoundly different from anything that is imaginable today. When William Smith was growing up, everything about them—whether they were commonly found examples like brachiopods or echinoids, whether ammonites or trilobites, gastropods or graptolites, or teeth or ribs or fragments of coral—was seen in a very different light. Assumptions were made about them and conclusions were drawn from their existence that bear little relation to what is today considered objective reality.

+

Pythagoras, it is often said, knew well what these mysterious bodies were, two thousand years before anything resembling the modern science of paleontology had begun shuffling out of the shadows. But, Pythagorean foresight aside, the world had long been steeped in a degree of ignorance that seems barely credible today.

Until the beginning of the eighteenth century the objects found inside rocks were known not as *fossils*—that word had a much more general usage, meaning anything, minerals and crystals included, that had been dug up from the ground. Any item that had been unearthed or discovered lying in a field and that had the look of an animal or a plant about it—an obvious shell, say, or a sea urchin, a leaf, or a piece of branch—was known, cumbrously though perhaps quite reasonably, as a "figured stone."

A few of these stones were easy to explain—some, like those that happened to have a shape vaguely resembling a human head, or a carrot, or a ship, had almost certainly been shaped accidentally. Tree limbs or animal bones that had never been mineralized and that were merely stuck in mud or in the sand by a riverbed were obviously pieces of modern organic life which had died and become mired in the earth. The figured stones that interested and amazed people in the seventeenth century—and people, aristocrats and members of the leisured classes especially, amassed enormous collections of them, with both the Royal Society and Oxford's Ashmolean Museum housing them in handsome display cases—were those that were clearly made of mineral material. These were thus definable as stones, and yet they looked uncannily like something that had once been living, or else they mimicked the aforesaid shells, sea urchins, leaves, or pieces of branch.

They obviously could not possibly *be* such things—that went without saying. To suggest otherwise was either to court ridicule—a once living shell, thrust halfway up a mountain, indeed!—or else to be accused of apostasy or heresy, for tinkering with the ordered faiths of nature. But to gaze at them in astonished rapture—this is what the nobly born of England did three centuries ago, much as later generations gazed in awe at mounted specimens of the coelacanth, or at specimens of rock from the surface of the moon.

No. Such things, so awesome and wondrous to behold, could only be explained in one way. Clearly they were unique creations of the Almighty himself—*lapides sui generis* is the phrase now employed ("stones unto themselves"). They existed for one reason only, and that was to reinforce in humankind's collective mind the omnipotence and imaginative beneficence of God. He placed the figured stones where they were discovered, using to do so what was termed a *vis plastica*, a plastic force. He used the force to insert into rocks miraculously perfect simulacra of living things, for the sole purpose of reminding the entire human race that God did indeed move in mysterious ways his wonders to perform. And there, to the enraptured viewers of the stones, was an end to it.

The science that was needed to justify such a belief to skeptics was simple enough. This, after all, was still the time of phlogiston* and the ether, and the firmly held belief that mountains grew like trees, organically, upward and outward. To anyone who imagined such a thing, it did not require too much of a leap of imaginative faith to conclude that mysterious stone objects found in the earth were there either because (*a*) they had been infused (on heaven's command) with some kind of petrifying fluid, (*b*) they had had their nature changed by a kind of juice that emanated from nearby mineral seams, or (*c*) that the stars had exerted some kind of magnetic or gravitational influence on them from the heavens. And if all these theories failed the rigorous tests of observation, then one could always simply resort to (*d*), the mysterious ways of God: Collectors would argue that a divine virtue was behind the placing of all fossils, using the word *virtue* in the old sense, rare now, of meaning "by way of supernatural power."

Old-fashioned scientific explanation appealed most of all, especially to those who, in post-Restoration England, were trying to make some order out of the chaos they perceived in the world. To the scientists the idea that a stone might grow into the shape of a sea urchin was surely not outlandish at all. If a perfectly symmetrical crystal could grow out of apparently nothing, if a mysterious process of chemistry could make a stalactite or a kidney stone or a coral—a rock that grows—then why could not the same kind of inexplicable and enigmatic natural force make a stone that looked like a shell, or a tree, or, as in the case of the Oxfordshire pound stones, in the shape of a hedgehog, and do so, moreover, deep within the body of a rock?

However, there was more to it than this. Even if the theoret-

*The first denial of the existence of phlogiston, the so-called "inflammable being" that was believed to be contained in all burnable objects, came with Lavoisier's discovery of oxygen in 1775. But throughout Smith's youth, phlogiston was the prime explanation behind flammability: Chemists only formally decided otherwise in 1800.

ical processes behind the formation of such figured stones were correctly guessed by these seventeenth-century philosophers, there was a host of additional unanswered questions: How did these figured stones get to all the places where they were found? Why did some kinds of rocks—those in wild moors of Devon, or in the mountains of North Wales, or the high hills of Shropshire—have almost no such stones buried within them, while other kinds, such as those that made the hills of Devon or were found in the quarries of Oxfordshire or the coalfields of Northumberland, possessed them in enormous numbers?

Why, as an early naturalist named John Rawthmell noticed in the 1730s, did most of these curious figured stones crop up inside those rocks that were to be found in a rough line that stretched in a northeasterly direction clear across England, from the cliffs of Dorset and via the Cotswold hills in the south, up through Leicestershire to Yorkshire and the great cliffs in the coast near Whitby?* And as corollary to this thought—if God was behind their distribution, why were the stones not left scattered around everywhere, to be found uniformly and randomly, like the stars?

It had been towards the end of the seventeenth century that the first very few and very bold observers raised (albeit timidly) the ultimate heretical thought: the possibility that perhaps, just perhaps, these objects actually *were* what collectors and scientists and countrymen had long been loath to consider admitting—the organic remains of the very creatures that they looked like.

It was men like Nicolaus Steno, a Dane, and Robert Hooke, a Briton, who blazed the trail: To them the unsayable became the irrefutable—these fossil stones, they were certain, had indeed

*I would never wish to dislodge my hero, William Smith, from his pedestal of honor as the father of English geology. But it has to be admitted that John Rawthmell's observation was more than prescient: One has only to glance at the modern geological map of Britain, or indeed at William Smith's own, to see how very right he was, and why.

once been living creatures.* Hooke argued his case particularly logically and meticulously. He identified three stages that could be witnessed on all sides, which he said demonstrated the three stages in the formation of a typical fossil.

In the first stage, wholly unpetrified bones, shells, and vegetable remains were to be found in beds of mud, peat, and moss. The rock around them was unformed, the fossils within still almost as organic as when they had been alive.

Then, second, in lignites and brown coals—the sedimentary beds that were not properly rocks but were slightly more solid and consolidated than mud and peat—there were bones, shells, and parts of trees and leaves that had been somehow *changed*. These specimens, which by now could perhaps formally be called fossils, had been half petrified. In their present-day resting place inside layers of half-formed rock, they too were half formed, being neither wholly organic, as when they were alive, nor yet wholly stone.

In the next stage they would become so. In layers of coal—a fully consolidated rock, though born from peat and lignite in turn—Hooke noted that there were leaf-, tree-, and other shell-like remains to be found that were as wholly coal-like, coal-colored, and self-evidently coal as coal itself. Could it perhaps be, he wondered, that great pressure, great heat, or complex physicochemical reactions had transformed the once organic remains into minerals, just as the mud had been transformed into peat, the peat into lignite, and the lignite into the solid black rock-mineral called coal? Could not a slow and uniform process, which had been so visible in the making of coal itself, work its mysterious magic on the life forms that had been present at the origin, turn them into stone, and make them into fossils?

*Although Hooke managed to avoid the strictures of the church for saying so, Nicolaus Steno, who published his ideas in 1669, was not so fortunate, being compelled by the dogmatic authority of the Copenhagen bishops to accept Ussher's unprovable notion that the world was 5,772 years old. He eventually gave up science altogether in disgust, and joined the church, the poacher remaining as gamekeeper until his death in 1689.

Most scientists of the time still dismissed such ideas as laughable. What event, they asked tangentially, could possibly have swept these remains to where they were now found? Could Noah's great flood (which was then implicitly and almost universally believed, as it would be for the better part of another century) have been so violent and so massive as to wash shells up onto mountaintops—where, it had to be admitted, they had been found? Could these creatures have been swept onto the land at the moment of Creation?

No to both, said the seers of the day: Noah's flood was said in Genesis to have been a short and placid affair, and as for Creation—since it was widely accepted that the land was created before life—it would be impossible for any organic remains to be infiltrated deep inside the newly created rocks because there was no life in existence to be so inserted.

In addition it had not escaped the notice of some collectors that many of the figured stones they found represented animals and plants that did not seem currently to exist. This suggested, in other words, that if indeed the stones were relics, they were relics of living creatures that were no longer around and had since become extinct. Since extinction was an impossible, unthinkable event in any divinely created cosmos, then this notion too was invalid, inappropriate, and wholly wrong.

✢

And yet, as the eighteenth century opened, so these long-held beliefs and prejudices were confronted with increasing vigor by counterargument, by solidly mounted challenges to the dogmas and received wisdoms and ecclesiastical imperatives of old, and, most important, by evidence.

The ideas of Steno and Hooke, however hostile their initial reception by the Church, however flaccid their initial acceptance by the public, began slowly to take root. At about the same time there came a vague, inexpressibly gossamer-fragile thought that there might be some kind of link between two of the concepts that were an implicit part of the fossil collector's system of belief. People began to wonder if these stones might actually be the relics of living things, and placed where they were found by no less an agency than what they liked to call the Noachian Deluge—Noah's flood.

Perhaps somehow the flood could be implicated in shifting these objects, even to where they now existed in the rocks of high mountain ranges and on the Oxfordshire meadows. Perhaps somehow this same flood could also be implicated in the process that created the objects in the first place. Perhaps the rocks and all that lay inside them—the Chedworth Buns, the pundibs, the oyster shells, the fern leaves, and the crystal corals, fish skulls, and lizard bones—had all somehow been precipitated or had crystallized themselves from the fluid of a universal, flood-created sea. Perhaps, if such things were demonstrably true, then maybe, just maybe, the matter of intense puzzlement that had already confused untold generations of naturalists—What were fossils and why were they found where they were?—might be solved.

The flood, in short, was to be the eighteenth-century answer to everything. Noah was now the key. Half a century before 1769, when William Smith was born, the notion that figured stones were just inorganic and petrified replicas, cunningly inserted inside rocks to prove the omnipotent genius of God, had been at last abandoned, conveniently forgotten, regarded if at all as a distant cosmic joke. A more modern and more reasonable science was on its way to being forged.

And if today the long survival of ideas about the flood, which must have colored and tainted the thinking of such an eighteenth-century observer as the young William Smith, seem more than a little ludicrous, then at least Smith was brought up free from having to believe that his pound stones and his pundibs were just minerals. He knew, as the thinking world was then coming to accept, that echinoids and terebratulids were not minerals at all, but, as Steno and Hooke had taught, had once been animals.

✠

So even though William Smith was brought up in a society still in the firm grip of purblind churchly certainty, his scientific training—such as it was—allowed for a measure of liberality. James Ussher was still there on the margins, to confuse; to deny his beliefs was to risk being branded a heretic. But in the later decades of the eighteenth century it was also possible, and moreover *acceptable*, for a thinking student to suppose that life, far older than humankind and perhaps far stranger than humankind could imagine, might once have existed on the planet.

The corollary to such thinking was that the earth must in turn be far, far older than James Ussher had supposed. That, for the time being, had to remain unsaid. But that it could be *thought*, and that there was evidence to prove it, was for the young Oxfordshire man, a liberating realization—a realization that helped in no uncertain manner to foster the new science that he was soon, and at first almost unwittingly, to help establish.

4

The Duke and the Baronet's Widow

Harpoceras falciferum

A fully equipped English duke, grumbled Lloyd George to what he knew would be a sympathetic working-class Edwardian audience on Tyneside, cost as much as two dreadnoughts, was every bit as great a terror, and lasted a great deal longer.

Which was not, it has to be said, an exact description of the third duke of Bridgewater. Francis Egerton, who was born in 1736, succeeded to the title when he was only twelve, gave much of his money and his collections to the government, and allowed the dukedom to die with him in 1803. He was a startling exception to Lloyd George's general argument, in other words: This particular dukedom of Bridgewater cost precious little and lasted almost no time at all.

Yet in one sense the prime minister was right, for Bridgewater was very much a terror—in many ways, seen from today's perspective, a quite appalling man. As a child he had been thought so stupid that his father, who was called Scroop, seriously considered making a codicil to his will to ensure that the boy, the second son, could never succeed to the title. The sudden and premature death of Scroop's oldest son, however, scuppered the plan—and the child, Francis, who would become a duke variously regarded as ignorant, awkward, and unruly, duly joined his fellow aristocrats in the House of Lords in 1748.

He was at first widely disliked. As a young man he was irredeemably philistine, with little regard for art or society. He dressed intolerably badly. He loathed flowers and all kinds of ornamentation. He smoked like industrial Manchester, consumed pounds of snuff, never wrote letters, and had arguments with everyone. Though in time he became a great collector of painting and sculpture—more for their value than for their beauty, critics sneered—he wasted little time on what he regarded as the fripperies of life. He was a curmudgeonly bachelor and a misogynist who so despised women that he would not even allow one to serve him at table. He had only two apparent interests—the racing and riding of horses and, most significantly for this story, the building of canals.*

It was a fascination that became an obsession, for both the duke and his country. "Canal mania," the national mood was called—and it was all begun by this strangely unpleasant man. In 1759 the duke of Bridgewater had completed a forty-two-mile stretch of artificial waterway, complete with locks, allowing him to ship coal from his own mines at Worsley, in Lancashire, directly into the heart of Manchester and then onward to the river Mersey. Since most of the price of coal was the cost of transporting it across country, the use of a canal slashed prices by as much as 50 percent. Smelling the prospect of limitless profits, every investor with spare change promptly jumped onto what seemed an unstoppable bandwagon.

Every bank, every entrepreneur, every developer, every engi-

---

*As it happens, the duke's cousin, the eighth earl of Bridgewater, is also connected with this story, though in a wholly unrelated way. He was a clergyman and a keen champion of the idea that humankind had been divinely created. In 1829 he left money to pay for a treatise that proved it: Geologists, many of them Smith's contemporaries and disciples, entered the contest.

neer in England suddenly seemed to believe that the canal was the highway of the future. The owners of the turnpike roads howled their dismay. Farmers, angry that their land would be torn up, raised all manner of objections. But, one by one, Parliament passed canal acts and navigation acts at a staggering rate. Small armies of navvies—workers on the inland navigations—descended on the hills and valleys to carve and cement these revolutionary new trade routes into place.

Grand plans were conceived for connecting the whole country, Carlisle to Cornwall, Dover to Dumfries, with a network of waterways. The great existing trade rivers of England, the Thames, the Severn, the Mersey, and the Trent, were all to be linked. Maybe, one overambitious inventor suggested, the English Midlands could have their own canal that followed the contour lines and so did not need the costly and cumbersome mechanism of locks. (Since this canal would have been hundreds of miles long, requiring a horse to drag its barge the equivalent of a transatlantic crossing merely to go down to London, the plan was quietly dropped.)

Almost overnight, extraordinary economic miracles were realized. A brewer in Burton who previously could reckon to be able to sell his ales within a radius of only five or ten miles, found he could now load his barrels onto a horse-drawn canal boat and two days later have them delivered in London. No longer did Josiah Wedgwood have to hear how his fragile porcelains had been smashed to smithereens during their transit on the potholed public roads; now they could pass along the waterways, in the steady tranquillity of the floating world, and be safely in the shops of Liverpool and Oxford and Edinburgh in a matter of days. Exporters based in Birmingham no longer needed to route all their wares through agents in London: They could send their goods to the United States directly, by canal boat from the factory straight to the clipper ships waiting at the docks. And for the ordinary public, too, canals became immediate sources of betterment: No longer did coal double in price in the aftermath of heavy rains—now it was always cheap, and except in times of thick, canal-choking ice, bad weather scarcely ever affected its price, or the speed of its delivery, again.

The duke was quite right to foresee that indeed in those early, heady days the greatest canal cargo of all was to be coal. One horse, plodding quietly along ahead of a fully laden coal barge, could haul eighty times more than if it were leading a wagon down a muddy road—could take four hundred times as much as a single pack-horse. All of a sudden anyone with a coal mine, anywhere in England, now wanted a canal—so that his anthracite and his steam coal could be carried quickly and cheaply to the furnaces of the Industrial Revolution.

✣

It has long been said that the people of England could never be poor, since they lived on an island made of coal and surrounded by fish. There had been an English coal-mining industry of sorts—via shafts and adits and opencut workings only—since the thirteenth century (though the Romans had known of coal and had probably burned it). From 1325 there is a record suggesting that a British mine exported a boatload of the strange black material to Pontoise, in northern France.

At first the black and flammable stones were used mainly for iron smelting and lime burning. It was only in Tudor times, when the climate turned chillier and demand for wood for house building soared, that people began to use coal to heat their homes. After that there was no stopping it. Wherever in the country coal was exposed on the surface—near Gateshead, close to Mansfield, outside Sheffield, in South Wales, near the Scottish town of Lanark—men clawed hungrily for it. It was convenient if the coal remained close to the surface: It was easy to work, and cheap. But it became so important a source of energy and heat that, by the fifteenth century, if a coal seam happened to plunge deep into the ground, then, discounting all risks in the name of profit, they promptly dug after it.

Coal miners were very limited at first. Mines flooded, they collapsed, noxious gases poisoned workers or burst into flame. But then came technologies that allowed miners to dig deeper, to pursue seams for longer, and as a result through the seventeenth and eighteenth centuries the industry advanced at a prodigious rate. Chain pumps were brought in from Germany, and mines became drier. Thomas Newcomen invented the atmospheric engine, allowing pits to go deeper, and allowing drowned mines to be pumped out and worked again.

At around the time of Smith's birth, as we have seen, James Watt came along with his condensing steam engine, and mines could be dug to reach seams four and five hundred feet deep; and then again a decade later, once Watt's double-acting steam engine had been perfected and its rocking beams had been adapted to move huge iron wheels, so everything changed. Air could be pumped down to the miners, water could be pumped from where it gathered, elevators could be created that would speed workers down to the coalface and that would haul them and their coal back up to the surface again.

✢

In 1800 all Britain's coal mines, in which men were now working as deep as a thousand feet below the surface, were producing a million tons of a variety of types of coal each year. Landowners realized that they possibly had beneath their lawns and meadows and forests huge seams of coal that could make them rich beyond their dreams. Everyone was suddenly on the lookout for dark rocks, for traces of blackness, for hints that somewhere below might be a lode of that rich, soft, sweet-smelling substance that was for England what emeralds and silver and diamonds were elsewhere. Pits were dug and quarries were clawed—but often recklessly, incautiously—at every spot where the earth seemed to offer up its dark temptation. More often than not the darkness was a chimera, a black shale, a slate, a mudstone, which had no more chance of burning than granite. Failures dogged the diggings of all too many countrymen: Some sort of guide, some sort of a *map* was needed, a way for men to forecast with some accuracy what might lie underneath them.

Men had been mining coal in northern Somerset since the thirteenth century—there is a cryptic reference in Roman writings to a house in Bath having been heated by such stone, locally mined. The Carboniferous Coal Measures that outcrop along the flanks of Pennine Hills in northern England, and in South Wales and southern Scotland, outcrop around the Bristol Channel too. The same hot dark swamps that eventually fossilize to produce coal existed south of Bath three hundred million years ago, just as they existed near Durham, Leeds, Mansfield, Lanark, and the Rhondda Valley and—since coal measures have been laid down all over Europe—just as they existed also in Silesia and Westphalia, in France and Belgium and across vast tracts of Russia.

The conditions in which they were formed were, miraculously for Europe's economic development, much the same everywhere. There were fetid and swampy jungles, all mud, dead ferns, and sagging branches of clubmosses and horsetail. The steamy, clammy air was thick with clouds of insects, including dragonflies as big as thrushes. Scorpions and millipedes scurried and squirmed among the grasses and primitive leathery trees. Amphibians—from large thickheaded crocodile-like beasts to more gentle salamanders—splashed and lumbered through the steaming pools.

But then, in a space of just a few hundred thousand years, and maybe less, the seas swished their way back, the trees and plants and animals were overrun by salt water and drowned and died, thin sands were laid down on top of all the dead vegetation, and then yet more limestone and shale and mudstone and marl formed and pressed down on the organic mat below, until all was hot and heavy enough for the heating and compression to begin—heating and compression that would turn all this thick, brown, decaying, gas-rich pulp into the hard, black rock we know today as bituminous coal.

In some places, where the world of the era three hundred million years ago was geologically stable—in Poland, say, in Westphalia, in northern England—the coal and sandstones that alternated with one another were thick and fat and relatively undistorted by any later tectonic events. But in Somerset and the rest of southwestern England and Wales, matters were very different. The layers of coal in the hills to the south of Bath have all been folded, closely, complicatedly, and very differently from the coal layers that are found in the fields of Poland or Nottinghamshire. The Somerset coals have been squashed into small, tightly wound folds and are fractured by countless faults and fissures—making them difficult to find, tricky to mine, and costly to pry from under the earth.

The crushing and twisting of the Somerset Coal Measures is evidence of part of one of the most dramatic events of more recent geological history. The rocks in this part of the world were all caught for millions of years in the gigantic vise of a cataclysmic mountain-building movement, one that occurred when the European and African tectonic plates of three hundred million years ago moved sharply and catastrophically against each other.

The grinding and squeezing and gnashing and crashing—basically, the closing of a huge sea called the Rheic Ocean that had divided Europe and North America on the one hand from Africa and South America on the other—went on for scores of millions of years, leading to the development of an entirely new supercontinent of the Permian period, called Pangea. The events, the vast rippling and crushing of the earth that so affected southwestern England, was once called the Hercynian orogeny. Now, like much in modern geology, it has a new name, the Variscan orogeny—and it has left a legacy of subterranean contortions and distortions that have greatly affected the appearance of all pre-Permian rocks of the region. It has not made too much of a difference to the scenery above—few phenomena above the surface in Somerset would prompt anyone to imagine millions of years of crushing and grinding. But what those years did to the underside of Somerset is truly awesome.

There are coal seams down there, but they have almost all fallen victim to the contortions of mountain building. The fact that the beds of coal have been crushed and distorted out of all recognition has had a profound effect on the local economy. It has not stopped miners from trying to pry the coal from beneath the ground—historically, very little dissuades them from that. But it has affected mightily the way in which the miners have over the centuries tried to do the prying. And it has made the actual process of mining very difficult indeed.

Coal that is difficult to obtain is priced accordingly—it is very expensive. And it was this simple fact—that Somerset coal was so very costly to extract—that led to the decision to build a canal. If it were costly to mine but cheap to transport, Somerset coal might be competitive still: A canal was essential to keep the coalfield in action at all.

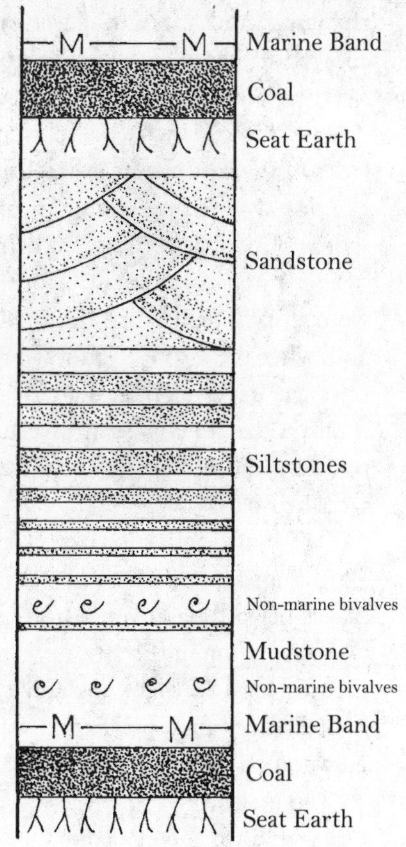

A cyclothem, or typical sequence of the rock types usually found close to a coal seam.

The fact that the tiny coalfield was wedged between two of the loveliest towns in southern England, Wells and Bath, was of little moment to the industrialists and profiteers who saw a chance to make money from the minerals below them. If this corner of rural England had to be made as charmless as Nottingham or County Durham, with the old stone villages defaced by winding gears and slag heaps, with the fields littered with cranes and coal barges and the air thick with smoke and the cry of steam whistles, then so be it. The English economy was changing, and fast, and it was imperative that the revolution be allowed to percolate into the slow-dropping peace of Somerset, for the good of all.

For it was a rich coalfield, and temptingly close to the big cities of the south, places like Bristol and Southampton and even London, where huge numbers of customers were now living and scores of new factories were being built. The coalfield's backers believed it had potential. More than one hundred tons of coal were already being torn from below northern Somerset in 1690; and when in 1763 a landowner named Lansdown decided to bore an experimental shaft near his village of Radstock, and found a sizable coal seam five hundred feet down, and another one six hundred feet lower still, there occurred the beginnings of a coal rush.

Suddenly villages like Camerton, Foxcote, Timsbury, Writhlington, and High Littleton, hitherto sleepy and forgotten places of wisteria-covered houses and fields with heavy-uddered cows drowsing in the afternoon sun, became crowded with burly men in hard leather caps and black smocks, and the sound of hammering and pickaxing and the irregular thudding of Newcomen steam engines displaced the music of skylarks and church bells.

It was a difficult coalfield, too. Not only had the Variscan orogeny wrought havoc with the seams—some of them even plunged vertically downward and could only be worked by the miners almost standing on one another's heads—but the roads in this part of Somerset were atrocious, thick with mud and as rough as the surface of the moon. John Wesley, the Methodist evangelist who once boasted that he regarded the whole world as his parish, found it difficult to include the North Somerset village of Midsomer Norton in his evangelical universe: It was so named, he wrote later, because the appalling local road conditions ensured it was only reachable in midsummer.

And then, in 1792, to make life even more difficult for the Somerset coal barons, Parliament was persuaded to pass the Monmouthshire Canal Act, which suddenly meant that South Wales coal, abundant and much simpler to work, could now become readily available on the Bristol market. This news resounded around Somerset like a death knell. The local collieries, it was feared, would be forced into ruin—unless, the owners decided, they also built a canal. If they could raise the funds and overcome the farmers' objections, then they could have Somerset coal floated swiftly and cheaply right into the heart of Bath. Perhaps, since another new canal was just then being planned to push deep into the southern heart of England to join the Thames, it could be barged right into the heart of London as well.

And thus did canal mania come to cider country. By February 1, 1794, after a committee meeting at the White Hart Hotel in Bath to decide the route—deciding which collieries would be favored, which ignored—a bill had been drafted and approved. By April 17 such was the anxiety of the promoters to beat off competition from the beastly Welsh that Parliament had passed it and King George III had signed it. The first excavators and the navvies then moved in a few weeks later, beginning the cutting, shoveling, and concreting that would end in the making of twenty-five miles of perhaps the least-remembered (and, as it happened, soonest-ruined) canals to be created during that curiously energetic period in British industrial history.

The promoters who first decided the route of the canal took their lead from the man they quickly appointed as official surveyor to the project—an absurdly young apprentice, lately arrived from Oxfordshire, named William Smith.

✣

The young man had come on impressively in the years since he was playing marbles with pundibs and marveling at the intricate beauty of pound stones. He had done tolerably well at school—though considering his family's poverty, there was no thought that he might go on to university. He had an apparent aptitude for geometry, he could draw more than adequately, and he had an evident fascination for the rocks among which he lived. His diary and his memoirs record his growing eagerness to understand what was going on beneath the green of the Oxfordshire meadowlands.

There are entries recording how he found the whiteness of chalk extraordinary, and how he wondered why there were no stones in the Churchill fields on which he could sharpen a knife or from which he could strike a spark. Notes tell how he collected crystals of fool's gold—iron pyrite—that workmen had found when they were draining a great pond in the village of Sarsden, that he had marveled at how some farmers were using a local blue clay to color their barn doors rather than waste money on paint, and that he had stood for many minutes enraptured in front of the earth-cutting machines being used to make a road through the Chiltern Hills, near Henley. And, unromantic though it may sound to the modern ear, he became fascinated by everything that had to do with drains, drainage, natural springs, culverts, bogs, and pools—a fascination that is easier to understand if it is remembered that to farmers in central England, earth, sun, and water were all—the core of their existence—quite proper subjects for their obsessive concern.

By the time William Smith left school he was something of a sophisticate, not least because, as well as having a developed rural knowledge, he had also traveled much farther than most young men from farms in rural Oxfordshire. He had gone many times to London. He would write later that he remembered especially sitting on a wooden seat on a cowhouse to watch criminals being hanged from the gibbet at Tyburn. He had been only too well aware of the scale of the terrible anti-Catholic Gordon Riots in 1780. He had experienced much of the Londoners' unparalleled joy when Admiral George Rodney returned in triumph after routing the French fleet off Dominica. And he remembered the thick mud stalling the horsecarts in what is now the eminently fashionable quarter of Manchester Square.

His uncle—who, because of his name, was known by his nephew as "old William"—seems to have treated him reasonably well, if somewhat parsimoniously: He was exasperated, it seems, with what he regarded as his nephew's effete habits—which by the time he was seventeen had extended beyond collecting brachiopods and echinoids to carving sundials in slabs of another local rock, the oolitic limestone of the Middle Jurassic.

Avuncular parsimony was nearly the boy's undoing. William found it difficult to afford books for his studies and had to go so far as to ask his uncle for an advance against his will—and which was apparently only given grudgingly—for a few shillings to buy the volumes he needed. (He had to travel to Oxford to make any serious purchases: The nearest town of any consequence, Chipping Norton, had no bookshop, as Smith would later note—"except for two shops that sold pots and pans and which sold spelling books.")

One of the volumes he did manage to acquire, however, was particularly well worthwhile: it was *The Art of Measuring*, by a man named Daniel Fenning,* and it was the book that introduced the young man to the skill that would become central to him for the rest of his adult life—the basic principles of surveying. And it was as he was carrying this book, walking down the

*Mr. Fenning was not knowingly related to one of the more notorious criminals of the day, a domestic servant named Elizabeth Fenning who allegedly poisoned her employers' family by serving them dumplings laced with arsenic. She was widely believed innocent, but was hanged anyway.

single sloping village street of Churchill, that he eventually met the man who would change his life.

His name was Edward Webb, and he was a professional surveyor. His craft was all of a sudden big business in England. Roads were being built, country estates being measured and laid out to gardens, canals were being dug, rivers improved—and common lands enclosed. It was the business of enclosure that had brought Webb to Churchill.

A group of the West Oxfordshire local squires and the wealthier farmers, just like their opposite numbers in countless other towns and villages up and down the country, had decided to have the local fields apportioned privately, and farmed efficiently. A surveyor was needed, and Webb was brought over from Stow-on-the-Wold, ten miles away. The young Smith introduced himself—in his own rather fanciful attempt at an autobiography penned many years later he wrote that he met Webb entirely by chance and asked him some penetrating questions about modern surveying practices. By the day's end, according to the diary Smith was now in the habit of keeping, today held in the library of the University Museum in Oxford, he had been hired to work as the assistant. This was the autumn of 1787. He was eighteen, and, informally educated though he may have been, he had a profession and a job.

It took him only a few months to master the basic skills. By the following spring he had learned how to use the pantograph and the theodolite, the dividers and the great steel chain. By the early summer of 1788 he was entrusted with doing his own work—the first opportunity arriving when one of Webb's older assistants, who, it was whispered, *drank*, miscalculated the area of some allotments he had surveyed and made their owners order fences of the wrong size. William Smith did the measuring and the mensuration all over again, got everything right, and was promptly set up by Webb to survey other tracts of land on his own.

He and Webb then began to travel together—indeed, Webb and his family so liked the young man they had him move away from Churchill and his niggardly uncle and into their substantial house at Stow. From there he traveled, to a farm in Cricklade, to make a survey of the Sapperton Canal tunnel, to the Braydon Forest, to the Kineton coalfield in Warwickshire. He also traveled to the New Forest, where he sank a borehole, looking for coal for the charcoal-burning industry, which was then locally booming.

And by chance he also made brief but memorable contact with the one other celebrated former inhabitant of Churchill—the former governor-general of India, Warren Hastings.

For his entire career Hastings, who had been born in Churchill, had yearned to return to his father's old estate at Daylesford, in Worcestershire, which his father had been compelled to sell because of what were delicately but opaquely described as "embarrassments arising out of the civil war" (between the seventeenth century's Royalist and Parliamentary armies, the Roundheads and the Cavaliers). With the eighty thousand pounds that he brought back from his thirty-four years in India in 1785, Hastings eventually managed to buy the great house—though at the very time he completed the purchase he was embroiled in the notorious impeachment trial (for alleged cruelty and corruption in Calcutta) that was to last for seven years, ruining him and (though he was acquitted on all charges) forcing his retirement from public life.

As soon as he bought the old family house and its 650 acres (for eleven thousand pounds), he decided that he needed the grounds landscaped. And in the spring of 1788 he called in the by-now-well-known surveyor Edward Webb from Stow, and his young partner-apprentice, his Churchill-born former neighbor, William Smith.

In his daybook Smith records meeting Hastings—"the gouty great man [who] sat on his horse with his livery servant behind him." The apprentice was rather harder on him than perhaps he needed to be: "Mr. Hastings," he wrote, "decided to be satisfied of my competency for the task, [and so] I had to sit down and draw him a sketch [of what was intended]. But in giving his instructions I was surprised at his ignorance of the scale to which he requested his maps to be enlarged." So, to add to the rigors of his trial before the House of Lords, the old governor-general now had a young whippersnapper of a neighbor arching an eyebrow at his poor reading of maps! The trials of public life, he may well have reflected, can be great indeed.

✢

This was, in general, an important time for William Smith—his fondness for travel, for the life of a gypsy rover, seems to have begun in earnest during the late summer and autumn of 1788; and with it—to judge by his diaries, which are forever noting the presence of this rock here and that rock there and the importance of this cliff or that valley, those rivers or that spring—a growing knowledge of and intimacy with the topography, and the wonder at what lay beneath it.

He was no great diarist; but once in a while his entries make one wish he had been a better one. His travels, and all that he saw during them, would have produced in more competent hand a glorious portrait of country life in late-eighteenth-century England:

> [A]fter crossing the naked hills from Stow, joyously with the thoughts of being trusted to survey an estate myself, I saw from the edge of Broadway Hill what appeared to me one of the most glorious sights in the world, and I well remember standing some time to gaze over the immense extent of the rich country below. The day was fine and the Vale of Evesham lay below me spread out like a map, the fruit trees and hedges being all whitened with the finest blossom that ever was

known. I advanced into this rich country to survey an estate at Inkborough, in the midst of apple trees. In the year before there had been a most abundant crop of fruit, so that cyder was exceedingly plentiful; and by the blossoms, another stock of that cheering beverage might be expected, for which the growers feared they could not find stowage, their casks being then all full. But before long these fears were dissipated, for the weather changed to wet and cold, the apples fell off before they had attained the size of walnuts, and the barley and other corn on the stiff lands I had to survey turned to yellow in the furrows and in all moist places. Most of the estate was upon the Red Marl which, in its redness, astonished me more than any other kind of soil I had seen.

He finally arrived in Somerset in 1791, on another mission from Edward Webb of Stow. Here he was to make a valuation survey, on his own, of an estate in the pretty village of Stowey that had recently been willed to a local grandee and coal mine owner named Lady Elizabeth Jones.* Characteristically, it being a warm and fine summer, the fares on the trans-Cotswold post-chaise being higher than it seemed prudent to pay, and, since a trek of fifty miles seemed to Smith no more than a casual stroll (as it would to most hardened geologists even now), he decided to walk.

He traveled along the roads the Romans built: After making southbound along the country roads he struck southwest along the remains of Akeman Street first, which took him from Burford down to Cirencester, and then turned on a more southerly route, via the Fosse Way to Tetbury and the old Roman settlement and pleasuredrome of Bath. After that, keep-

*Lady Jones, the widow of the baronet Sir William Jones, was the great-grandmother of the astonishing Angela Burdett-Coutts, who was a great patron of geology, and who endowed two lectureships in the science at Oxford.

ing to the hills to the south and west around the spa, he walked by way of Radstock, Odd Down, Stoneaston, and Temple Cloud—coalfield villages all—until, finally, he reached the vast acreages of the Jones estate.

He was to remain in this part of England, at first working for his patron and then later for the Somerset Coal Canal Company, for the next eight years—eight years during which he would make the discovery, come to the realization, announce the deduction—and begin the hard grind—that would earn him his place in posterity.

5

A Light in the Underworld

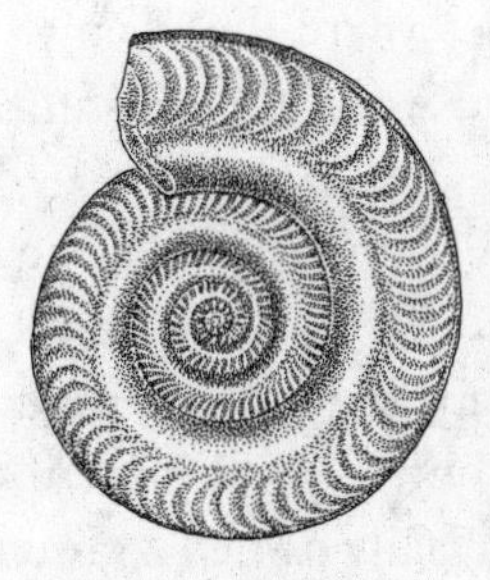

Hildoceras bifrons

The village of High Littleton is a gritty, windy, hilly, and generally unlovely place, almost as far removed from the traditional imagery of Somerset—no cider apples, no jolly farmers, no thatch or maypole or cricket on the green—as if it were near Nottingham or on the country fringes of industrial Lancashire.

Much the same might be said of the entire swath of countryside that lies ten miles to the south and southwest of Bath. It is pinioned between some of the loveliest and most measured architecture in all of southern England. Wells Cathedral, with its statues and gargoyles, is the spiritual and aesthetic mooring point to the west. The ordered, bewigged, and precious life, the powdered and pomaded air of England's once-second city, Regency Bath, wafts in from the east. But between lies an area that still today looks curiously out of place, architecturally, atmospherically, socially, commercially: It has smaller houses, meaner shops, grubbier streets, a spoiled and ragged landscape.

All is a direct consequence of geology. These fifty square

miles or so of Somerset, bounded by the red-brick villages of Clutton in the west and Combe Hay in the east, Priston in the north and Kilmersdon in the south, lie on top of a score of complex, broken, twisted, and contorted seams of coal, which until as late as the 1970s were worked by as independent and militant a band of English mining men as might ever have stepped out from beneath the winding gears of the coalfields of Durham or Lanark. Maybe their militancy had arisen because of the unusual proximity, in these parts, of their class enemies—all around them the great limestone houses and mannered city terraces were occupied by soft-handed gentleman farmers and sportsmen, philosophers and squires, artists and divines. There was no other local industry to provide brotherly support: In the fifty square miles of country that unrolled itself around where the twenty-two-year-old William Smith came to live and work, the laboring classes were coal miners, to a man.

The mines were owned privately, usually by whoever owned the land under which the coal was first found. Of the Lady Jones who first employed William Smith to landscape the estate around her house in Stowey we know very little; but she and her late husband, Sir William Jones of Ramsbury, certainly owned a great deal of land other than the immediate neighborhood of her house, which included both a large number of the neighboring coal mines and a collection of farmhouses. It was into one of these, Rugborne, which stood on the eastern outskirts of High Littleton, that she allowed her new young employee to move.

Smith had good reason to remember this house vividly. It was, he wrote, a large old manor house, three-storied, solid, and foursquare, sheltered by lime trees, and with a walled courtyard in front, and steps leading up to handsome gates and walls (now gone) that were thick enough to have a series of rounded niches in which Smith liked to sit and study his books. The house had once belonged to a Major Britton, who, the locals said, had ruined himself financially by working the coal seams that ran beneath the house. But when Smith moved in it was occupied by a tenant farmer, Cornelius Harris, who gave him board and lodgings for half a guinea* a week, and took in his horse for an extra ten shillings a month.

He remembered the house not so much for its architecture or its comfort or the eminently reasonable price of the accommodations, however. He remembered it because of the work that he was to engage in, first in a mine less than a mile away to the north, and later in the canal that he was to help build a little farther afield. It was work of staggering significance.

And because of it, Rugborne Farm, High Littleton, deserves a memorial. The work that Smith undertook there, and the results that he achieved and pored over there, led ultimately to the creation of an entirely new science. For years afterward he looked back on his time at Rugborne as the most important in his life, and the house as the crucible of the new discipline he believed himself to have created. "The birthplace of geology," he later said, grandly. But there is no memorial—not a plaque on the house, nothing—just an incorrectly dated sign erected by the local council back in the thirties, at the entrance to a lane, pointing halfheartedly to where the house still stands.†

He worked there first for Lady Jones—surveying, planning, draining—in her capacity as director of the High Littleton Coal Company. He was not without a greater ambition: A letter found in the files showed that he tried hard to persuade her to allow

---

*A guinea, equivalent to a pound and a shilling, is a classically British and very informal unit of currency—with neither a coin nor a bill to formalize it—that is still used today (despite Britain's having adopted decimal currency in 1971) in some circles, such as the buying and selling of racehorses and sheep. There used to be a one-guinea coin, struck from gold from the eponymous nation, but only its name and worth survive, and today the word is only a vague and ephemeral throwback to more casual financial times.

†Smith's obsessive interest in cartography rarely left him, and his jottings give an indication of how his mind was working when first he came to Rugborne. "[M]y residence was most singular, it being nearer to three cities than any other place in Britain: it is 10 miles from Bath, 10 from Bristol and 12 from Wells."

Rugborne Farm, Smith's first true home near High Littleton, which he called "the birthplace of geology."

him to become a shareholder in one of her newer mines, and to be its general manager. There is no record of a response. Instead he was compelled to work at one of her older pits, which had been first excavated in 1783 and which was to have nearly thirty-five years of working existence. The mine was called the Mearns Pit, and though it was certainly one of the less familiar and less prosperous of the hundreds of mines and shafts that had been sunk over the centuries into the coal measures of North Somerset, its importance on the global stage is quite inestimable.

For the Mearns Pit at High Littleton has a standing in the history of geology that is comparable to the one that Gregor Mendel's Moravian pea garden has in the science of genetics, the Galápagos Islands in evolutionary theory, and the University of Chicago football stadium in the story of nuclear fission. Yet this

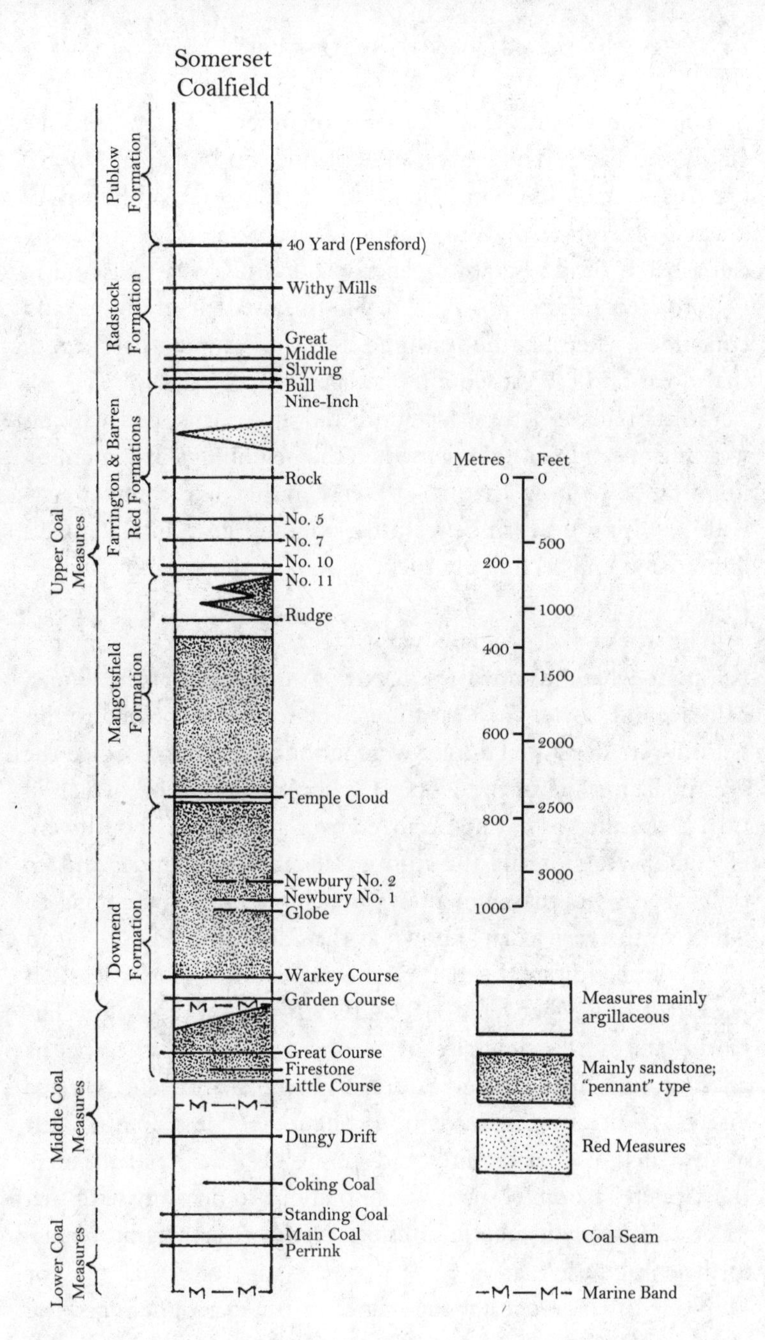

A section of the Mearns Pit in Somerset, where Smith first noticed the succession of rock and fossil types.

Somerset coal mine also goes unremembered today, just like Rugborne Farm. There is no blue plaque, no brass plate, nothing. Just a small lump in a field that marks a scarified hillock of grassed-over mining waste, a mound that Somerset people still call a batch. And a scrap of paper in the University Museum in Oxford, recording what William Smith saw, felt, thought, and concluded when first he ventured into the dripping darkness of that dreary little West Country colliery.

To describe exactly what Smith did at Mearns, and why his working and thinking there marked the beginnings of something important, we now have to descend, rather like a miner nervously waiting his turn beneath the winding gear, into the half-illuminated and technically tricky gloom of the scientific underworld.

The rocks that contain winnable coal in North Somerset belong to what are known, perhaps not surprisingly, as the Upper, Middle, and Lower Coal Measures. The rocks that belong to this period were deposited during what in today's Britain is called the Westphalian stage of the Upper Carboniferous period, which in turn is classified as having occurred over a period of about twenty million years during the closing billennia—the upper end, in other words—of that immensely long chunk of geologic time (an "era," in the vernacular) known as the Paleozoic.

Modern radiometric dating shows that the Coal Measures were laid down between 310 and 290 million years ago. The world had by this time already seen and welcomed the appearance of such living complexities as ferns, sharks, and winged insects; it already disposed, by extinction, of less complex life forms, such as stromatolites and graptolites. Before the end of the Upper Carboniferous it was also about to do away with that most attractively lovable lobsterlike Paleozoic arthropod known as the trilobite.

None of this—neither the names of the geological eras, nor the age of the rocks, nor the existence of a fraction of the life forms within them—was known to Smith or to anyone who worked with or around him. When he climbed down into the dark mine on a cool summer's day in 1792 he knew just about nothing about the sorts and conditions below—nothing except that there was definitely coal in these hills, and the miners who clawed it out of the ground knew only too well how it was organized, and how to make some kind of sense of the chaotic state in which it existed.

Several shafts made up the Mearns Colliery, and Smith went down every one of them—clambering down slippery ladders and ropes or being taken down with the chain of one of the dredgers that was powered by a Newcomen steam engine. He would write that he looked closely at the walls of the shaft, and noticed the succession of layers of rock as they slipped past his eyes.

At first he was rather puzzled. "The stratification* of the stones struck me as something very uncommon," he was to write a short while later,

> and till I learned the technical terms of the strata and made a subterranean journey or two, I could not conceive a clear idea of what seemed so familiar to the colliers. But when these difficulties were surmounted and an intelligent bailiff accompanied me, I was much pleased with my peregrinations below, and soon learned enough of the order of strata to describe on a plan the manner of working the coal in the lands I was then surveying.

His paper plan—the first he is known to have written, and preserved in the great Gothic cathedral-museum in Oxford—is merely a scrap, titled "Original Sketch and Observations of my

*This appears to be the first time that William Smith uses a term deriving from the word *strata*, the study of which would so dominate his life as to become his nickname: To all nineteenth-century England he would be simply Strata Smith. The *OED* suggests that the words *stratum* and *strata*, meaning a layer or layers of sedimentary rock, became current in England at the end of the seventeenth century; Smith himself was the first to use *stratigraphical* in 1817; *stratification* made its first appearance in 1795.

First Subterranean Survey of Mearns Colliery in the Parish of High Littleton." It records, without comment, what he saw of the technology that was employed to bring the coal up from below. Horses that were harnessed to a windlass lifted the coal from the three-bushel carts on a tramway "that are wound up the second gugg [an underground incline], drawn along here to the bottom of the first by a man down on his hands and feet, bare, with a cord round his waist, to which is fastened a chain that comes between his legs and hooks to the forepart of the cart."

Then he went down into the pit and noted exactly what he saw. First, there was the grass and gravel of the surface, which blended seamlessly into topsoil, eight or ten inches thick. Next came a more solid rock—at first broken into small chunks, then progressively more solid, and generally red in color, though with layers that were strangely green. It was, so far as he could make out, an earthy limestone, a marlstone, interspersed with shales—the marlstone being very red, the shales having a green tint. Overall, though, it was red and earthy—and similar, so far as he could see, to other red marls, other red earths, that he had seen around Bath on his walk there from Oxfordshire.

Then he noticed something very important. These marls, he wrote, were not laid down horizontally—but seemed to slope very gradually away to the east. They had a dip, an angle from the horizontal, that seemed to Smith to be about three degrees. They also had a strike, which is the compass direction of the line drawn where the dip intersects with the horizontal, and which in this case was aligned about 95 degrees from due north. The marl beds were sloping, in other words, in the general direction of London, and if their strike continued on the same general heading, in the direction also of Europe. Whether they continued to slope away in this direction and at this rate, and whether the marls were thus eventually buried deep below the capital or beneath the hills of France, Smith could neither tell nor imagine. But what was obvious to Smith is that they pointed in that general direction, and that they did so on a very gentle downgrade.

The chain-rope was clattering him slowly downward, and toward an important moment—for as he continued to gaze at the slowly passing walls, so everything around him on the sides of the mineshaft abruptly changed. Within a matter of inches the nature of the rocks—their look and feel and color and hardness—was all altered.

Where there had been these limey, shaly, reddish-greenish marly limestone rocks above, now instead on the sides of the shaft was a facies of thick, sandy, grayish brown rocks that appeared wholly and profoundly different.

Moreover, if he looked closely at the gradation between one type and the other, Smith could see that this change did not even occur within a matter of inches—it was much less, so much less that in fact there was a verifiable point at which the change occurred, a line above which was one rock, the Red Marl, and below which another, the Grey Sandstone, without any gradation between them that was worthy of the name.

And it wasn't just the nature of the rocks, their lithology. Just as noticeably the *attitudes* of these new rocks had changed too: Where the Red Marls had sloped neatly at a gentle angle, all pointing in one direction, these new gray-brown sandstones—ugly and rather unappealingly harsh rocks, as they appeared to Smith in his guttering candlelight—appeared to plunge deeply downward. Their bedding planes, the cracks that separated one bed of sandstones from another, were steeply inclined, and in places they were very evidently folded, and in places so folded they had been shattered and broken up by dozens of small dislocations, as though the bands of rock had once been contorted by a giant vise, all twisted, flattened, and cracked, and then released and left to lie tumbled and in disarray.

It was all so very strange. Back up at the top of the shaft, near the lip, had been these beguiling, neatly laid-out sheets of almost horizontal strata, with just their faint slope in the vague direction of London. Now, below them, the rocks were contorted and thrust in all manner of different directions, most of them dipping down toward the center of the earth. As mentioned in the previous chapter, modern structural geologists using radiometric dating techniques, taking deep-drill cores, and performing seismic examinations have recently established that Somerset's coal measures were folded and twisted by the great Variscan mountain-building period that occurred 290 million years ago, when the African and European tectonic plates collided. But none of this was known or even vaguely conceived by Smith or by any geologist then alive. So Smith could only stare at the junction between the rocks and wonder—Why? How? How could one possibly make sense of such a bizarre arrangement?

✠

By chance this was exactly the time when the stripling science of geology was indeed trying to make sense of such structural oddities. James Hutton, the gently born Scots doctor who was one of the leading philosophers of the age, was at the time of Smith's work just three years away from publishing his seminally important three-volume book *The Theory of the Earth*, which sought to explain countless facts about the earth—including precisely how such an internal discordance among strata, a so-called unconformity, had first come about.

It was a very similar unconformity, in fact, that had first set Hutton thinking. He had seen another arrangement of rocks, much like those Smith was seeing at High Littleton, in a rocky cliff on the North Sea coast of Scotland, at a now-hallowed site called Siccar Point in Berwickshire. There, thick layers of the Old Red Sandstone of Devonian times were—and still are, of course (though now much chipped by the geological hammers of ten thousand students)—lying flat on top of steeply inclined hard gray sandstones of the Silurian period. Hutton suggested that both of these sets of rocks must have been laid down by deposition from a sea, and that the gray sandstones, after being laid down, had been lifted up and out of the sea by great and mysterious crustal processes, had been subjected to all manner of contortion and twisting and deformation, then eroded so that the top of all these twisted rocks was cut off and smoothed and flattened, and finally submerged again by the sea in Devonian times to allow red sandstones—what would eventually be called the Old Red Sandstone—to be laid down on top.

And the important ideas that Hutton grasped from pondering this, and which he declared in the final volume of the great book, were these: first, that the processes of formation of rocks are the same processes—creation, erosion, deposition, change, and erosion—that we see today, whether in a stream bed, by the sea, or on the flanks of a volcano; second, that the earth works today more or less as it always has; and third, that these processes, tiny and unimportant though they may seem in the short term, take place over unimaginably long periods of time, over an uncountably large number of years, so that anything that is observable in the rocks around us today can be imagined as having happened as a result of normal, everyday occurrences taking place over aeons of geological time. "The present," said Hutton most memorably, "is the key to the past."

If one could imagine this, wrote Hutton—and he wrote these words at almost the same time as Smith was clambering down into the Somerset mines—then one could think of the earth as having "no vestige of a beginning and no prospect of an end." It was a staggering thought, quite appalling to some—but it was the kind of thought that was to permeate the world of geology into which Smith-the-surveyor was just now making his entrance.*

---

*In Korea there has long been a tacit recognition that small earthly processes, carried out over millions of years, can in the end have a geologically significant result. There is in Korean mythology a famous measuring unit that denotes a very long period of time. To gauge how long that period is, one is asked to imagine a mountain, made of solid granite, exactly one mile high. Once every thousand years an angel flies down from heaven and brushes the summit of the hill with her wings. The unit of time represents the number of years it would take for the angel and her summit-brushing wing to erode the mountain down to sea level. Given long enough, of course, she would do it. As would a stream, or even the wind—providing that geological time was encompassing enough—and was far, far longer than the mere six millennia allowed by Bishop Ussher.

✠

It was in and among the gray sandstones that the true wealth of the Mearns collieries, and all the neighboring mines, was held. For these were coal measures, and in them, as Smith was to see as the mine chain clanked him lower still, were—thin, distorted, and plunging deep below—the precious veins of coal.

The miners, tough and taciturn men, had given names to all the seams, as miners all around the world always do. Coal seams may all look the same, much as all coal may look alike; but in fact all seams as are different as all people, when they are looked at as closely as a miner has to look. Each band has an internal pattern, a collection of fossil types, an appearance, a feel, a *personality*, that makes it instantly recognizable. A miner will be lowered down the shaft and step out into the blackness and know in an instant where he is, how deep in a series, how far down within a seam.

✠

Some of the less romantic workers gave their seams mere numbers—Newbury No. 2, or simply No. 7 or No. 11. Others were named for obvious reasons—the Great Course, the Little Course, the Standing Coal, the Main Coal, the Nine Inch, the Coking Coal. And then there were the seams with stranger names, no longer easily explicable: the Slyving, the Dungy Drift, the Globe, the Perrink, and the Kingswood Toad. *Go work the Slyving today*, might be the order of the morning. *Try your luck with the Perrink*, or *climb down into the Nine Inch, if it's not too wet afoot.*

The men who first found those seams and named them—after their favorite pub, perhaps, or a girlfriend, or a hill immediately above the mine—are long dead now, and such histories as have been written of the coalfield give the Mearns Pit, one of the lesser collieries in commercial terms, short shrift indeed. But when the last truckload of coking coal left the last mine in Radstock for the Portishead Power Station in 1973, it was from one of the deep seams of the field, nine thousand linear feet in absolute terms* below the red earth that lay serenely and unconformably above the contortions of the coal.

The stratigraphical order in which the different types of rock were arranged in the coalfield, as the local miners knew and as William Smith learned from them all too rapidly, had an utterly predictable regularity to it. Smith would see and come to know the strata intimately as he saw them one by one, again and again, as the great winding chain lowered him still further down through the measures.

First there would be the sandstones themselves, with strange patterns of bedding about them. Some patterns were large and reminiscent of dunes, other patterns were small, like the ripples to be found on the beds of river estuaries. After a few dozen feet, the sandstones would peter out and be gradually replaced by argillaceous beds of more thinly bedded rocks, with finer, darker grains—muddy and silty, solidified versions of what one might find on the bed of a shallow river. Next there would be a bed with fragments of shells, recognizable as the kind of bivalves, *Carbonicola* and *Anthracosia* among them, that are found in rivers and abhor the dissolved salts of the sea.

Then there would be mudstones—softer, dark, friable rocks, suggestive of—well—solidified mud. Next, an indicative precursor of what would come next, there would invariably be a thin band of lighter-colored, harder rock stuffed with shells of a saltwater brachiopod like *Lingula*—a creature still to be found more

*A lot less than 9,000 vertical feet in terms miners would understand, since the coalfields were set on their edge, or otherwise deformed, by the Variscan orogeny. The deepest pit in 1792 was a mere 450 feet; and when the mines closed in 1973, the deepest was the Braysdown Colliery, 1,700 feet. To get to the lower level in a coalfield where the seams are so up-ended, the mines had to be sunk in the right place, and so did not have to be very deep.

or less unaltered today, which lives today as Carboniferous *Lingula* did 290 million years ago, happily and resolutely among the salt and the wavelets on the seashore.

And below that—the coal. The seam might be the Globe or the Perrink or the Kingswood Toad, and whether it was nine inches or nine feet thick, it would always be overlain by the same facies and underlain by the same seat earth. However, and most important for the ever-observant William Smith, each seam would be subtly different: It might be colored richly and deeply black, it might be oily, flammable, stiff with flies and footprints and the relics of eminently burnable fossil vegetation—*Mariopteris*, *Annularia*, *Lepidodendron*—and then, suddenly, cut off in a matter of fractions of an inch, the whole black band would be underlain by a light band of nonflammable seat earth, in which once all the ferns, trees, and horsetails placed their roots and from which they gained their sustenance.

The pattern, Smith saw, was always the same, in mine after mine after mine: from top to bottom, Sandstone, Siltstone, Mudstone, Nonmarine Band, Marine Band, Coal, Seat Earth, and then again Sandstone, Siltstone, Mudstone, on and on. On top of everything, placid and unconformable, the red marls, the flatly sloping beds of the startlingly red red earth.

And through all of this, in mine after mine, in quarry after quarry, what was perhaps the most crucial realization of all in this time of early, primitive discovery—the fact that recognizable seams of coal would always be in the same position compared to one another. The Dungy Drift—identifiable by its thickness, its color, and its fauna and flora—was always above the Perrink. The Rudge was always above the Temple Cloud. Never once—unless the Variscan folding has turned the whole bedding upside down, and a skilled observer could always tell if a rock bed was the right way up or not—would the Slyving be anywhere other than above the Great Course and the Firestone. It would always be well below the Withy Mills Seam and what the miners had christened the Pensford Forty Yard.

And that, in essence, is what William Smith first learned, both from the miners that he met and from the sections of the underworld that he saw. He understood for the first time that geology was a science requiring observations in three dimensions. He could make maps and make surveys of the visible upper dimensions of the landscape with ease—anyone with a modicum of a skill could do that. But to see *below* the surface, to observe or

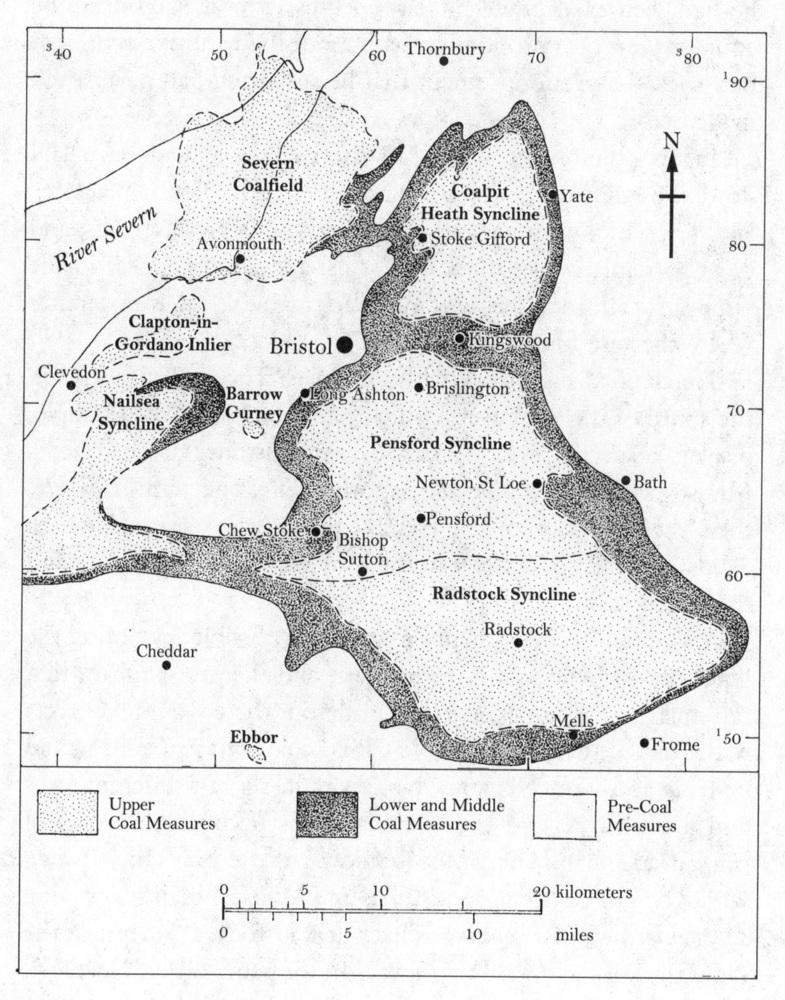

The Somerset Coalfield.

extrapolate the imaginable third dimension underground—that was a new skill, possessed by very few, and yet that had a potential that Smith was soon going to recognize and exploit.

But first he had to grasp and set down on paper the importance of what he had just seen. He pondered on what he remembered; he pored over his notes. And he promptly wrote down in his diary what it was that he realized—that to judge from what he had seen at all these pits around his farm at Rugborne, he could make a firm pronouncement about the nature of sedimentary rocks—a pronouncement that he suspected had never been made before.

In his opinion, he wrote, all the rocks that had been laid down as sediments at a particular time in a particular place are laid down in a way that has much the same characteristics, and most particularly just the same fossils, and always appear in the same vertical order, in the same stratigraphical order, no matter where they are found.

He clarified his theory with examples. The coal measures in the south of the Somerset coalfield, near Radstock, for example, display a specific order. There is a Sandstone, a Siltstone, a Mudstone with *Carbonicola*, another Siltstone with *Lingula*, then the four feet of the Temple Cloud coal seam, back to Sandstone and Siltstones and then the Newbury No.2 seam and the Globe seam.

Using Smith's new theory it would be possible, by noting the types of fossils he found, to forecast, and then to confirm, that ten miles to the north, at High Littleton, there would be exactly the same order. The Temple Cloud coal with its fossils would be lying above the Newbury No.2 with its slightly different ferns and fossil leaves, and the Globe with its peculiar collection of once-dead inhabitants would be down at the base. In between would be just the same Sandstones and Mudstones, just the same *Carbonicolae* and *Lingulae* as back at Radstock. Perhaps not the same thickness. But always, always in the same order. Never—if the beds were lying right side up—would the Globe be at the top, never would the *Lingula* be found above the *Carbonicola*.

Such an order of strata would be repeated also in other places that had never yet been explored. The order would be repeated also in mines yet undug. It was an order that could be well and accurately predicted. The fossils would be the key to working out what the order was. Using them, one could forecast the precise succession of the beds underground. And if they could be forecast, they could and would eventually be mapped.

✣

This was true for the coal mines of High Littleton—of that much William Smith was now certain. Yet at the same time as he was realizing and understanding all this, Smith began to wonder: If what he had found was true for the seams, facies, and lithologies of all the rocks that he and the miners had found lying below the red earth—might it not also be true for all the rocks, for the Limestones, Oolites, Shales, Clays, Cherts, Marls, Sandstones, and Silts, that lay above it? And might it not be equally true, too, for rocks so far unfound, and which would presumably lie underneath the coal? Was not this predictability of strata likely to be a universal phenomenon?

The miners said no. Smith records their instant rejection of his theory matter-of-factly. "The order of superposition in the Coal Measures at each pit seemed well enough known to the colliers," he wrote in his diary,

> and on drawing a section thereof with nine veins of coal I was naturally led to ask whether the superincumbent strata, rising into hills two hundred to three hundred feet above the mouths of their coalpits, were not also regular. I was told there was "nothing regular above the Red Ground," which in their sinkings varied much in thickness. This did not deter me from pursuing my own thoughts about this subject.

It was just as well that he was not deterred. He thought about the miners a little more closely. He felt he could understand why, out of a mixture of protection and plain ignorance, a miner might insist that these particular patterns and fingerprints of rock successions that Smith had recognized were confined to *their own* rocks, to the coal measures of the Upper Carboniferous, and would not be reproducible elsewhere. He might understand the miners' motives—but what they said made no sense at all.

Wasn't it more likely that some similarly arranged succession of strata was actually to be found among *all* the rocks of England, whether they were above or below the coal? Whether they were younger or older than it? And further, wasn't it likely, if this orderliness of succession proved true elsewhere, that someone with a good eye and a good imagination could find the arrangements and the possibilities for identifying and following unseen strata, the hidden underground strata, among all the rocks of the world?

Might there not thus be some way of predicting what lay where, how deep it lay, how thick the beds were likely to be, and what might lie above and below it? And thus, might there not be a way of drawing a guide to this hitherto hidden underneath of the planet, in much the same way one drew guides to the visible world, to the simple topography of the overburden?

Had he not, in thinking so, stumbled onto an original, fundamental truth? Wasn't it likely that everything he had reasoned for the rocks at High Littleton was true for everywhere else as well? And if it was, then wasn't it likely that everything geological, everywhere—whether it was underground or overground, whether it was deep or shallow, whether it was visible or not—could be predicted, could be drawn, and thus could be mapped?

✠

New observations were needed. More data, more facts, more work, below and beyond the very special world of the coal mines, beyond the age-limiting, fossil-limiting, lithology-limiting purlieus of the Carboniferous. William Smith needed a bigger canvas on which to sketch the first portrait of what he was now nervously beginning to imagine.

And it was then, thanks to connections, location, and coincidence, that William Smith stumbled onto the chance that made him. The coal from Somerset's dozens of mines needed to be moved. The perfidious Welsh across the Avon, having caught the duke of Bridgewater's fierce mania, were reported to be building a canal and getting ready to move their coal along it—and suddenly all Somerset was fretful, its miners and mineowners concerned that the county might lose out to Wales, that its coal would never get to the markets. A great Somerset canal urgently needed to be built.

William Smith, who was by now an established master of all the local mysteries of coal, a clear and present friend to the local landed mineowners, and known to be clever with the theodolite, the plane table, and the chain, was the ideal man to be involved. He knew how to carry out a survey. He was obviously the man to plan the route, to make sure the canal snaked properly from coalfield to market. William Smith, it was decided, should be the Somerset Coal Canal's first surveyor.

He accepted the job with almost unseemly relish. He had a motive that he never vouchsafed to his new employers. Not only were the wages excellent, the perks more than acceptable, and the possibilities of share options in the new canal tempting—but the process of building a canal meant that, quite simply, a great swath of the county needed to be sliced open, cut neatly and deliberately in half.

And in the process of cutting the land he might be able to confirm his theories, and see if that original and fundamental truth was indeed a truth at all. By slicing open this vast line of survey, and then building a deep canal halfway across the county, the land itself would for the first time be exposed. It would be laid bare and fresh for Smith to see, to examine in detail, and to wonder if he might, just might, be right.

Right in thinking, that is, that one could tell which strata were which by their nature and by their enclosed fossils. If one could do that one could, in theory, find and identify the outcrop of a particular stratum in one place, and then find and identify it in another place and another, and before long be able to draw a map, from which it would then be possible to extrapolate, with accuracy and speed, the position of that stratum as it snaked through the entire English underworld.

One could do it for one stratum or, with patience, for all strata. One could then draw a map of the underneath of England just as readily as one could map the overground. And if it might be possible to map the underneath of England, then by extension one could make a map of the hidden underside of the whole wide world beyond.

It all depended, though, on his making one so-far-unmade discovery: He needed to find that the aspects of rocks that were so recognizable within the patterns of the coal measures, occurred just as well in the rocks that lay above them. The miners were skeptical. But William Smith was not. He believed that there would be a pattern out there. He needed simply to lay open a great slice of English countryside and see for himself, firsthand. The new canal would be his one opportunity for doing so.

6

The Slicing of Somerset

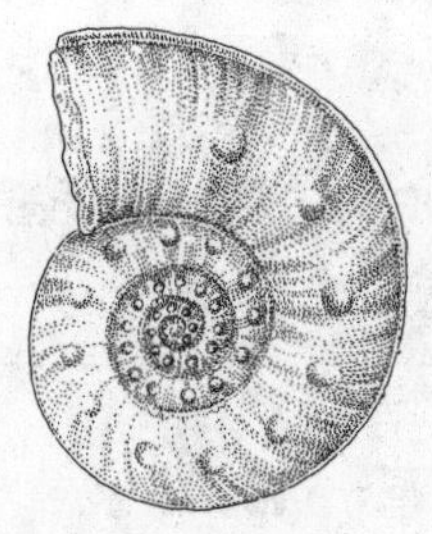

Sonninia sowerbyi

The British have an unrequited love affair with their railways. The older, the more obscure, the smokier, the more inefficient, and less commercially successful they are, the better. Dr. Richard Beeching, whose infamous 1965 report resulted in the closure of five thousand miles of old, inefficient but much-loved track and the attendant two thousand railway stations—most of them wrongly remembered as cottagelike and fretworked, with endlessly congenial stationmasters and rose beds planted on the platforms—is still regarded as a villain. The evidence of Beeching's savagery—abandoned lines now swathed in grass, old bridges rising over emptiness, stations now turned into houses or small factories, or left to rot—remains everywhere. And whole communities in remote and pretty parts of Dorset, Cumberland, Norfolk, and Yorkshire curse him yet, as the man who ruined forever an enchanting and supremely British way of life, along the country railway.

The Camerton & Limpley Stoke Railway, in North Somerset, was as pretty a railway as they come. It was known by local schoolboys, and for obvious onomatopoeic reasons, as the Clank. Its economics, however, made no sense at all, right from the moment it opened for business in 1907. Its tiny income—from a dwindling number of coal mines, from a mill that packaged wool dust, and from the carrying of luggage to and from a boys' school—doomed it to extinction even before Lord Beeching had the opportunity of getting his hands on its seven miles and seventy-eight chains of track. The last fare-paying passenger traveled on the morning after Valentine's Day, 1951.

But the Clank was memorialized in the minds of many million of Britons of my generation because it starred, though unrecognized by most who saw it, in one of the most successful British films of the time. It was called *The Titfield Thunderbolt*, and it was a comedy, made in 1952. It told the story of a line that was due for closure but might be awarded a reprieve if it could show that it could be run, by the villagers who depended on it,

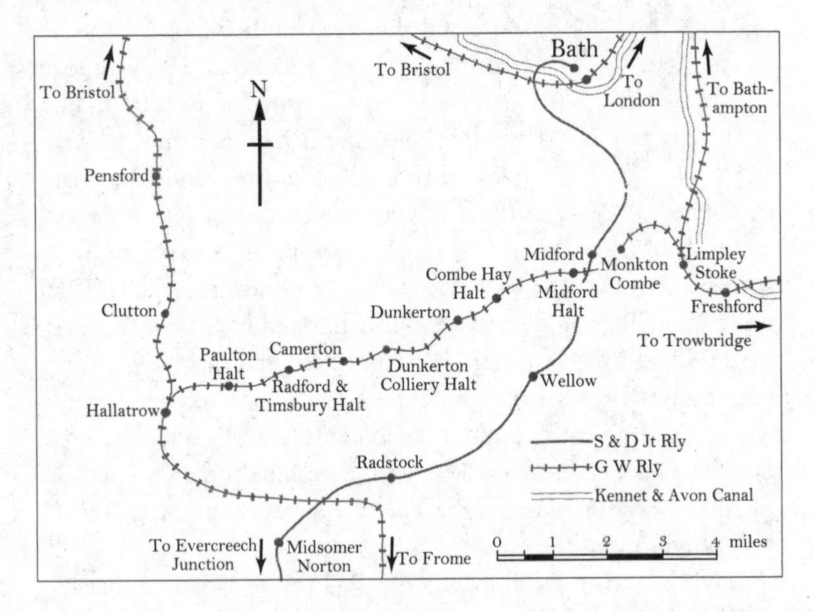

The Camerton & Limpley Stoke Railway.

with greater efficiency than a competing local bus service. The train was run by a team that included the vicar, the local squire, and the ladies of the Women's Institute. The bus, by contrast, was owned by a pair of curmudgeonly and profiteering blackguards from a grim slum town nearby. Who won and who lost I will leave for those who have not seen the film; but for this account of William Smith's life, the story of the film is less important than the setting in which it was made.

For *The Titfield Thunderbolt* was filmed in the valleys of the Cam and Midford Brooks, at the eastern end of the Camerton & Limpley Stoke Railway, in countryside that was—and still is—as lovely and as unmistakably English as any landscape imaginable. The film seemed then, and still seems in its time-warped look today, to be set in the middle of some kind of utterly English Elysian fields, where all is sun and lush meadows, babbling brooks and thatched cottages, village greens and cricket matches. On all sides there are comfortable pubs and ample barmaids; the people are by and large sturdy and honorable. The soundtrack drips with a fine nostalgia: There is birdsong, and there are steam whistles, we hear a milk churn being loaded, the flap of a porter's flag, distant peals of church bells, the lowing of dairy-ready cattle, and behind it all, as bass continuo, the amiable chuff of steam engines as they amble through cuttings and over level crossings and bustle back down the valleys to their sidings and their home.

But this is no fantasy of an imagined Englishness. The railway may have gone, but the world in which the *Thunderbolt* used to run is still there, south of Bath. It has been preserved in some kind of Betjemanesque amber—a patchwork of landscape six miles long by three miles deep, between the river Avon in the east and the village of Combe Hay, halfway westward along the long-disused railway line.

But its beauty peters out very quickly, and with sudden drama. To the west of Combe Hay the land becomes much less interesting, less pretty. A passerby in the train, were it still running, would—if traveling westbound—notice the change most easily, would see how the rural idyll between Limpley and Combe becomes slowly more tinged and tainted by the first indication of industry, of smoke, grit, iron, and rust. By the time the engine reaches Dunkerton, a couple of miles on, the smell of coal dust hangs in the air, and by the next station, Dunkerton Colliery Halt, there is (or was—it has long been demolished) the winding gear of a mine. And then from there to the west all is coal, all is industry, all is grim. It takes a small effort of imagination to recall that only ten miles back down the line, back to the east of Combe Hay, there was pretty landscape—landscape of a loveliness from another world.

The reason, as so often, is the geology. The hills around Combe Hay and Midford Halt, by Midford and Limpley Stoke itself, are the outcrops of what is called Bath stone, a warm, honey-colored oolitic limestone of the Middle Jurassic. A reporter for the *Somerset Guardian* understood this well when, in May 1910, he wrote of a railway journey that "there is not a more prettily situated line in the locality of Bath . . . the run through the Oolite from Combe Hay to Monkton Combe is the most interesting part of the track, because the traveller has lovely views all the time."

The oolitic limestone dips gently eastward, much as did the red marls that Smith found in the Mearns Colliery. What this meant to a traveler heading west on the Camerton & Limpley Stoke Line—the map shows it passing in an almost direct westerly direction for most of its route toward the terminus at Camerton and the junction at Hallatrow—is that he or she would pass—or chuff or clank—steadily downward through the geological table, because of the steady dip of the rocks. From the start at Limpley Stoke station he or she would pass much of the way through the Jurassic, from Middle to Lower. Somewhere around Combe Hay Halt he or she might have noticed having entered the outcrop of Triassic rocks. By the time the train has reached Dunkerton Colliery, the traveler will be in the thick of the Upper Carboniferous, and of the coal.

This much we know today, and a great deal more besides. In William Smith's time, however, very little was known—and anyone who made that westbound journey from Limpley Stoke to Camerton in 1792 might well have marveled at the change of scenery but would have had precious little understanding of which rock was which; which type might be older or younger than any other; and which appeared where, when it did, and why.

Anyone, that is, except for William Smith. For seven seminal years these few square miles of gently graduated English loveliness were to become Smith's stamping ground. He worked in precisely the area along which the Camerton & Limpley Stoke Railway ran for the 44 years of its commercial existence. He did so because, 120 years before the railway was built, he was to become, after only the briefest of apprenticeships, the man responsible for surveying the canal—a canal that would provide ready-made the route that the railway itself would later take. As it happened, the railway was built to compete with and ultimately replace and ruin the old canal. But here this matters little: What is important for an understanding of Smith's work is the decisions that were taken by him as to the canal's initial route.

Both the canal and railway had perforce to start at the same place—Limpley Stoke—because that was the junction for the bigger canal (the Kennet and Avon, along which goods could go to Bristol or to London) and the main-line railway (along which goods could also be taken to the same two industrial centers). Both ended at the same place, Camerton, because that is where the coal was. But the precise route that was taken between these two end points was, essentially, up to William Smith to decide.

The process of choosing that route was to offer him an intimacy with countryside and landscape that was never to leave him. And it was to set him wondering, too, about all those mysteries that eluded, or did not even appear to concern, those others who might travel between the two ends of the coal mine route. Why such a journey began in an area of limitless beauty, and why it ended in a region so very much less attractive, would to them be either an enigma or a matter of no consequence.

But not to Smith. He was different; his view was different. He alone would in time come to recognize that the simple gradation in the rural loveliness of the canal route said something well worth knowing about Somerset's mysterious underworld. His genius—the unanticipated genius of this uneducated farmer's son—was that he realized it was not simply a matter of noticing the difference. It was also possible—desirable, and perhaps important—to find out just *why* there was a difference in the first place.

His survey of the canal was the means to such a discovery. In a sense the fact that he was making a new canal became eventually almost incidental to his own self-allotted main task—which was to find out why the landscape was the way it appeared to be, and whether any of the lessons he had learned in the coal mines, and which the miners insisted belonged to mines alone, could apply out in the wider geological world as well. The red marls of the High Littleton mines dipped east; the oolite and Triassic rocks of the canal route dipped east—so could any firm prediction that he made about the one be equally applied to the other? Smith thought so; and the survey would confirm or not, as the case might be.

In making the route for the new canal he would be digging his way through the very rocks that made the hills—lovely but unproductive hills in the east, their aesthetically unremarkable but richly endowed equivalents in the west—that stood in the way of himself, of the canal, and of progress. If only what he found would ultimately confirm what he had suspected from his explorations in the shafts at the Mearns Colliery in High Littleton, then his work for the newly formed and comfortably subscribed Somerset Coal Canal Company Ltd., was likely to be of earth-changing importance. Smith knew that if he cracked the code he suspected he might find during his surveys, then in time he could become a famous man.

For William Smith was now a changed and changing figure.

Until he moved down from Gloucestershire to Somerset he was a man of seemingly modest vision. The small epiphany that occurred during his stay in High Littleton showed him the advantages of ambition; once he had started to work for the canal company that new ambition was to be annealed and case-hardened, until Smith became convinced that he would one day, and with good reason, enjoy a place in history.

He needed a brief period of apprenticeship. By good fortune the renowned Scotsman John Rennie—a towering figure of the day, a man who specialized in making the massive, in building lasting structures like dockyards, bridges, tunnels, breakwaters, and lighthouses—was working nearby. Rennie had evidently heard talk in the local inns of the parliamentary petition for a new small canal in Somerset, and, always eager for new commissions and fresh work, he signed up to make the initial survey for the route of what was first to be called the Dunkerton & Radstock Canal. But he was too busy to work alone, and needed help.

Two members of the Somerset canal committee, to whom Lady Jones had enthused about William Smith's acuity and intelligence, suggested his name. Rennie agreed to meet Smith, and liked him immediately. The great engineer hired the young surveyor on the spot. It was a moment that changed Smith's life forever—particularly since Rennie's own idea was that, if all worked out well, Smith himself would in short order take on the job of surveyor and engineer for the entire canal project. And this is precisely what happened: He got the full-time job, and embarked on an association with the Somerset Coal Canal that remains central to his reputation to this day. Seek out any local enthusiast who can still discern the old ruined waterway snaking along its forgotten route through the hayfields: *Smith built that, you know. Great man.*

The young man threw himself into the job with great enthusiasm. Not only did his new responsibilities allow him to rub shoulders with such notables as John Rennie and another acclaimed canal builder who was also working nearby, William

Jessop. Not only did it allow him to explore a particularly lovely piece of English countryside. Not only did it give him an opportunity for both the advancement of a budding career and to take part in the creation of a monument. It allowed him also, and at last, to test his grand ideas.

Armed with his theodolite and his chain, his trenching tools and his shovels, Smith thus began the long process of slicing through Somerset. Each day he would venture out from Rugborne on his horse, making his way slowly and methodically a few hundred yards each day, from the mines of Camerton eastward toward the thatch of Limpley Stoke. He would set up his theodolite and his marker pole and his compass and his chain, he would measure out a section, he would write his notes about what he found and would speculate on the projected ease or difficulty of digging a canal along the path he had chosen; and then he would pack up all his kit and, piling it onto the back of his long-suffering nag, would move forward another few hundred feet and begin it all again.

Unwittingly he managed to create in his design for the route of the canal a device that would help him enormously in his geological inquiries. He and his colleague engineers decided that for most of the route west of Midford there should actually be *two* near-parallel canals—there should be a northerly route, the Dunkerton Line, and a more southerly, the Radstock Line. By this arrangement of double-branching the canal, the coals from the entire coalfield could be collected with a minimum of effort, and funnelled eastwards towards the junction with the Kennet & Avon Canal just by the huge aqueduct at Dundas.

The arrangement made good sense from the coal-barons' point of view. It also made good forensic sense for William Smith—for it meant that whatever observations he might make in one branch of the canal, he could check in another branch that ran only a mile or so away, in an almost precisely parallel direction. He had always wanted Somerset to be sliced open: Now it

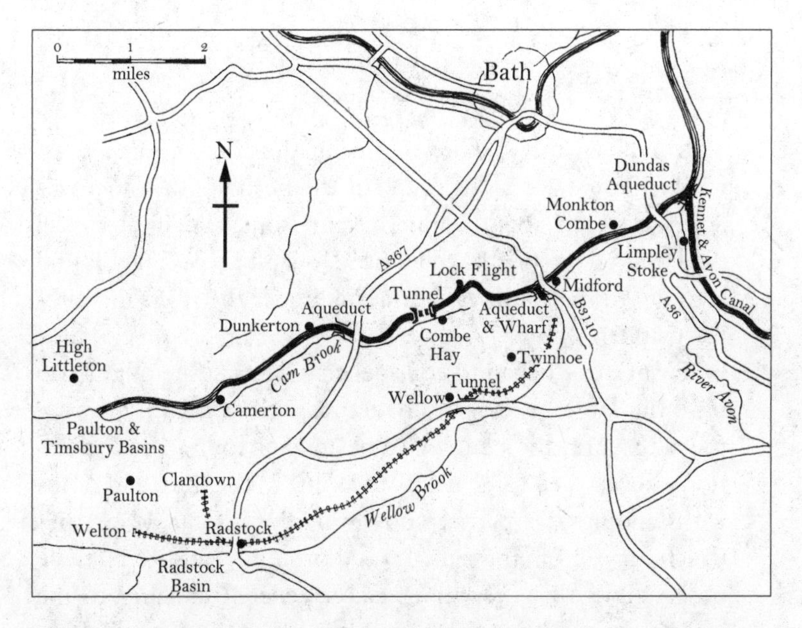

The Somerset Coal Canal, showing the two near-parallel branches that later helped Smith in confirming his theories.

was being sliced and filleted twice over, and all his work could be compared, checked, and confirmed.

It was as he worked patiently forward in this fashion that he first noticed something. He was surveying a stretch of the proposed northern route, between Dunkerton and Midford, when he saw exactly what he had been hoping for:

"I observed a variation of the strata on the same line of level, and found that the Lias* rock which about three miles back was a full 300 feet *above* this line was now 30 feet *below* it, and

*The Lias—the name is variously thought to be Old French, or a Cornish quarriers' term for "layers"—is a bluish, highly fossiliferous argillaceous limestone, used in building and the making of tombstones as it is hard and takes a high polish. It occurs in the geological table at the lower end of the Jurassic, just above the Triassic. Although the respective ages of the formations were unknown to Smith, he will have seen that the Lias occurs above what he and his collier colleagues called the red earth, or the red marl, and which in turn lies unconformably above the coal measures.

became the bed of a river, and did not appear any more at the surface," he wrote.

> This induced me to note the inclination of the same rock, which I knew was to be found at the head of two other valleys lying each about a mile distant from, and in a parallel direction to the one just described*—and accordingly found it to dip to the south-east, and sink under the rivers in a similar manner.
>
> From this I began to consider that other strata might also have the same general inclination as well as this. By tracing them through the country some miles I found the inclination of every bed to be nearly the same as the Lias; and notwithstanding the partial and local dips of many quarries which varied from this rule, I was thoroughly satisfied by these observations that everything had a general tendency to the south-east and that there could be none of these beds to the north-west.

This, it can be argued, was Smith's most momentous early realization. It was an utterly simple observation—that when going from Camerton to Limpley Stoke southeastward along a line that was both continuously straight in direction and consistently horizontal in attitude, the old rocks that he had been accustomed to seeing in Dunkerton fell away and then vanished beneath his feet, while as he approached Midford newer and newer rocks appeared in the cliffs and cuttings before him, only to fall away and vanish and be replaced in their turn. These layers of rock, he said, were arranged "to resemble, on a large scale, the ordinary appearance of superposed slices of bread and butter."

---

*By this he meant the more southerly arm of the new canal, the so-called Radstock Line—although it is something of a mystery that he refers to "two valleys," when the Radstock Canal surely had only one.

He took notes, and he drew sketches—none of them too clear today, none of them too confident. Those who look at his drawings fancy that they see narrow shaded bands that they imagine mark the outcrop of the coal measures and their unconformable relationship with the Triassic rocks above—and one would-be biographer thinks this might fairly be called the first true geological map. But it isn't, not really—and all one can truly say of the cartography Smith did during his canal days was that he was most definitely thinking about it. He began, he wrote in his diary, "to delineate on maps the courses of the strata, and constantly traced and retraced the order in which they would be intersected in making the canal." His greatest achievement at this stage was not that he drew any maps—that would all come later—but that he noticed things, drew conclusions, and laid plans. He became uncannily able to perceive the spatial geometry of the world beneath his feet—to imagine, on the basis of what he saw above ground, just what the world looked like underneath.

Today some critics remark sniffily that all this came too easily. That Smith was merely lucky, in that his chosen layers of rock all had a uniform dip to them, and they varied along their outcrop in interesting and obvious ways—he could with no difficulty recognize how the limestone turned to clay and then to shale, to oolite and so on—and so the layers revealed themselves to him quite plainly. Some also say that other surveyors had noticed much the same thing many years before Smith did—an amateur member of the local gentry named John Strachey, it is argued, who worked in much the same part of the world, spoke of the layers of rock looking like "the leaves of a paper-book."*

Some also point out, quite fairly, that Smith came at first to

---

*Strachey, in his *Observations of the Different Strata of Earths and Minerals*, written in 1727, speculated that the layering of the rocks in Somerset had been caused by their being rolled up by the rotation of the earth.

some naively rash conclusions about the easterly dip of the rocks he examined. It occurred to him, for instance, that since the oolites and the Lias and the Trias and the chalk of southwestern England had this convenient attitude, then probably *all the rocks in the world* were arranged likewise. The strata, he wrote, "form a set of lines extending from Pole to Pole with a regular inclination to the East. And the motion of the earth, which probably commenced while these strata were in a soft state or of a pulpy consistency, would naturally place them in an inclined curvilineal position."

He realized before long that he had misjudged this badly; and he made a host of other errors besides, with which his critics make hay. But William Smith was the one man who took his observations and formulated his idea and wove the whole cloth of new theory from it. Maybe the local geology did make his task relatively easy. Maybe he wasn't the first to notice the arrangement of the local rocks. But he saw the rocks, he made a deduction—and he then took that deduction to its logical and, as we shall see, its astonishingly beautiful conclusion.

✣

After six months of heavy labor in the river valleys south of Bath, with the initial survey almost done, Smith's employers suddenly decided he should interrupt his work and take part in a brief expedition. Before the first sod was dug and the first layer of stone laid—matters that were in any case the business of engineers, not surveyors—Smith was to go off on tour and take an instructive look at how other people were routing and then building canals. So in the late summer of 1794, along with two members of the canal committee, Samborne Palmer and Dr. Richard Perkins, William Smith was instructed to set out in a post-chaise on a journey that would last two months and would take him on an expedition over more than nine hundred miles of England and Wales.

He now knew that the colliers' rules for coal, and his own ideas from High Littleton, applied more generally to the hills and valleys he had surveyed elsewhere in Somerset. Was there any chance that they applied, as he thought they might, to the nation as a whole? In the Rugborne stable yard, as the postilion saddled up the horses and Smith clambered up into the cart alongside the driver while Palmer and Perkins sat in the carriage behind, he must have felt a welling sense of apprehension, of the excitement of his confident certainty. He already knew he could draw a map of a coal mine. His surveying work showed he could draw a map of North Somerset. Now, as the horses struck sparks on the coachyard cobblestones, he knew he might soon have the opportunity to do the same for England.

7

The View from York Minster

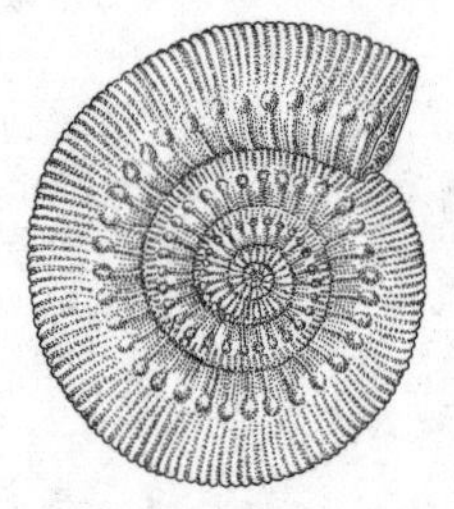

Stephanoceras humphriesianum

His excitement and his apprehension were still very much a secret. For the first few days in the coach he vowed to say nothing to his traveling companions. William Smith, all of a sudden, was worried.

Earlier in the year he had become gripped by an anxiety that is all too common among discoverers who work alone—an anxiety that must have been even more frequent in times when communication was so restricted. For although he was by now quite confident that his assumptions about the order and regularity of stratification were right, he was frantically concerned in case others might be thinking along the same lines.

He had no intention of announcing his ideas to Palmer and Perkins. He would not in fact for the moment disclose his theories to anyone. He was still not wholly certain he was right, and he had no intention of blurting out an unformed set of ideas and thereby making a fool of himself. Nor did he want to claim as his own notions that might rightly belong to others—and yet at the same time he hoped against all hope that he had originated these ideas, that he truly was the first to think them.

Shortly before setting off on the trip he had managed to dispel at least some of these doubts. The canal committee had traveled to London en masse to give evidence before Parliament—a bill was needed before the construction of any canal could begin, and though the questioning tended to be perfunctory, MPs needed to know that all was in order, the money had been raised, all opponents duly heard, all process duly done. William Smith was a witness, and recalled with some awe that, even though he was only twenty-five, he was summoned to the bar of the House, sworn in by bewigged jurists, and compelled to play a significant part in the democratic proceedings of the time.

But aside from a day or so giving testimony, he had little else to do. All his expenses were paid, and he was on salaried commission with precious few duties to perform until sent back to Somerset. So he decided he would go burrowing and scavenging through the libraries and bookshops of the capital, searching for anything that might show him what he feared—that like-minded others were on the case, that in creating the still-unchristened science of stratigraphy, he was not entirely alone.

We have no firm idea of what he found. There was, however, a lot on offer. We do know he acquired at about this time some basic reference books—Plot on *Oxfordshire* and again on *Staffordshire*,* John Morton on *Northamptonshire*, and John Woodward's majestic *Catalogue of English Fossils*. We know he was assiduous in his searches, and that he came home with a considerable library.

Almost certainly he would have come across John Strachey's *Observations on the Different Strata of Earths and Minerals*, written in 1727, and with the engaging theory that all rocks are spinning when laid down, and so are revealed as outcrops as they solidify, their edges like an unrolled sheaf of papers. Maybe he

*This is the same Robert Plot after whom the Chedworth Bun, or milkmaids' pound stone, *Clypeus ploti*, was named. See chapter 3.

saw John Woodward's other great work, his *Natural History of the Earth*, published in 1695, which revealed his fascination for "speluncae, grottoes and wells," and his unshakable belief that "the circumstances of these things in remote countries were very much the same as those of ours here."

He would perhaps have gained considerable insight had he found a copy of *An Inquiry into the Original State and Formation of the Earth* by John Whitehurst, which was popular enough to have two editions, in 1778 and 1786. Though the book put great emphasis in Derbyshire (still one of the counties most useful to teachers of geology, and so swarming each year with students, equipped with magnifying glass, compass, perhaps a now-environmentally-incorrect bottle of hydrochloric acid, and an Estwing hammer), it laid out a more general program that Smith himself would follow, almost to the letter:

> It is my intention to have deposited specimens of each stratum, with its productions, in the British Museum, arranged in the same order above each other as they are in the earth; being persuaded that such a plan would convey a more perfect idea of the subterraneous geography, and of the various bodies enclosed in the earth, than words or lines could possibly express.

But we have little clear idea of his reaction to anything he discovered. Would Whitehurst have inspired him right away? Or would he have been depressed on reading it, suspecting that he had rivals? Would he have been dejected to read Christopher Packe's 1743 description of the geology of Kent, his so-called Philosophico-chorographical Chart, since the author insisted that his was "a *real* scheme, taken on the spot with patience and diligence." Or by *The Course and Phenomena of Earthquakes*, published in 1760 by the Reverend John Michell, who took Strachey's rolled-up-papers view of the world and,

> bending them up together into a ridge in the middle, conceive[d] them to be reduced again to a level surface, by a plane so passing through them as to cut off all the part that has been raised; let the middle now be raised again a little, and this will be a representation of most, if not all large tracts of mountainous countries.

Would he have been dejected by such thoughts and such writings? Intimidated? Apparently not in the slightest. Indeed, he seemed almost smug. "Although several authors had noticed the thickness of some strata in succession in various parts of the country, their resemblance to others was never noticed—none were collated, and *for want of comparisons there could not be any reasonings on the subject* [emphasis added]." By the time he had left London and was back at Rugborne readying himself to set out for the Bath post-chaise terminus, he was displaying a remarkable equanimity. No such observations as he had made, he wrote, had ever been placed on record. At least—not yet, and not in England.

Yet as the party trotted off northward from Bath, he still steadfastly kept his own counsel. He was, he would later write, *overjoyed* to be going on the journey. It would go everywhere he wanted—they would pass northeastward through the English Midlands up to Leeds and York, along the east coast up to Newcastle-upon-Tyne (where there were no planned canals—none north of Leeds, in fact). Then, after a brief halt in an inn, they would cross back over the Pennines and head home via Lancashire and Shropshire and the borders with Wales. Perkins and Palmer sat inside the carriage, chatting endlessly about the possibilities for profit: They were interested only in using the tour to find out more about the mining and carrying of coal—Samborne Palmer was a landowner and suspected his estates contained abundant riches, if only they could be transported to market swiftly and cheaply.

But Smith, though professionally interested in all matters flammable, had grander designs. He sat invariably out in the open, perched beside the driver and his blunderbuss-equipped guard* constantly scanning the horizon, continually asking to be allowed to get down and flail away at some roadside exposure "with the small hammer he seemed always to keep with him," as his nephew was later to write, and to bring specimens of rocks, fossils, crystal, and minerals back into the coach with him.

"The slow driving up the steep hills," he noted in his diary,

> afforded me distinct views of the nature of the rocks. Rushy pastures on the slopes of the hills, the rivulets and kinds of trees all aided in defining the intermediate clays. And while occasionally walking to bridges, locks and other works on the lines of the canal, more particular observations could be made.

Outwardly he was there for Perkins, Palmer, and the Somerset Canal Company. But "the most important [of my interests] I pursued unknown to them; though I was continually talking about rocks and other strata, they seemed not desirous of knowing the guiding principles."

More than likely the couple thought of him as a fusty old bore, and laughed at him a little from behind their Woodstock gloves. But it has to remembered too that people, especially those older than Smith and the small army of similarly curious and inventive younger people who were coming to the fore in the England of these times, were likely to be as much perplexed as wearily amused.

Their world that had seemed so stable for so long was now changing all too rapidly, and men like Palmer and Perkins only half understood what was happening. They might have recog-

---

*Highwaymen were a menace during the period, and travelers encountered them so frequently that women were advised to carry two purses, one for the robber, one for themselves. The short-barreled blunderbuss, invented in Germany, was a wildly inaccurate weapon, but highly effective at disposing of a highwayman at close range.

nized in their strange companion what some of today's middle-aged recognize in the young electronics visionaries—that Smith was a man who, though part of their world, still had a view that was somehow much larger than theirs, that he had firm sight of a future that he somehow *knew* was better, as well as being a future that was definably different and, most crucially, utterly unlike the world of the present. William Smith knew that he stood on the edge of something; and that knowledge, that certainty, set him somehow apart and made other, more ordinary men uneasy.

They crossed the Cotswolds, put down at Tetbury, and again at the head of the river Thames, and took time (as Smith had already done with Edward Webb) to see the huge canal tunnel at Sapperton (because the initial plans for the coal canal called for a long tunnel to be built near Combe Hay). They looked at the Kingsnorth Tunnel on the Worcester & Birmingham Canal—Palmer remarking disdainfully (according to Smith's diary) that the young surveyor spent an unduly long time inspecting the walls of the tunnels, gazing intently at the rocks through which they had been bored, hammering bits from the sides, and taking away fossils and samples until the coach groaned under their weight.

They passed on through Derby and Ripley, dropped in to look at the great palace of the dukes of Devonshire at Chatsworth, went on to Matlock; and at Lord Fitzwilliam's mine at Hisley Wood, Smith and Dr. Perkins were lowered deep into the pit in a basket, suspended as at High Littleton from a quintainlike crossbar worked by a steam engine. And then at Leeds, at the outer edge of the Yorkshire coalfield, from which point north there were no more canals, they decided to transmute themselves into tourists and visit the three cathedral towns of York, Durham, and Newcastle, and thereafter call it a day.

Some like to say that it was high in the tower of York Minster that Smith first broke his silence and told his traveling companions about his bold ideas. There is little hard evidence for this, but it makes for a compelling tale—written most recently as fiction in a remarkable book, *The Floating Egg*, by a keenly original Yorkshire geologist named Roger Osborne.

According to his account the three men climb the 290 steps from the nave to the top of the tower—Smith bounding ahead, the others panting damply and having to pause to catch their breath on the transept roof. When finally they reach the top, where there is a view for thirty miles in each direction, they find the young Smith standing transfixed as if in a dream, bright eyed with obvious excitement, caught in the headlights of a remarkable idea. They ask him what excites him.

"You see those hills, Mr. Palmer?"

"I do."

"You may not know it, but it is possible from their contours, the lack of trees, and their general appearance to say they are made of chalk.'

"That is a remarkable skill. I congratulate you—"

"No, no, that is not the point." His impatience with me was perhaps necessary and I tolerated it. "It is the pattern that is everything."

"I see," I said, though I confess I did not.

"You remember the canal at Kingsnorth, and the locks to the north of Birmingham?" He was animated now, as if excited himself by the words he was saying.

"I remember them well."

"The rock they passed through was a kind of red marl and sandstone mixed up."

"I remember you saying so at the time."

"In each case it lay as unconformable cover to the coal measures."

"What exactly does that mean?"

"It lies over them, but there is an interruption between the beds, that is all. Just as at the colliery at High Littlejohn [*sic*], and all over Somerset."

"I see."

"And then we came on to the limestone of the Lias, in Derbyshire and here in Yorkshire, just as in the Cotteswold Hills."

I must have looked a little confused at this, as Mr. Smith saw the need to explain further. He grasped my hands and held them flat and horizontal between his.

"The rock strata are formed like this, the oldest beneath and the youngest on top."

"Yes, I know that much." I might have been annoyed at him, but his enthusiasm was a pleasant tonic.

"Now, everywhere we have been, the rocks have tended to dip like this—" he twisted our hands slightly "—toward the south-east."

"I understand."

"So that when the top is levelled off, at the surface of the ground—" Smith slid our hands in their diagonal aspect so that the edges of them made a flat surface "—the oldest rocks are to the north-west and the youngest to the south-east."

We both looked at our hands for a moment or two; then became a little embarrassed and dropped them. Mr. Smith looked out across the country, and seemed in danger of re-entering his private world.

"And the chalk, Mr. Smith?"

"Ah yes, the chalk, Mr. Palmer. The chalk, you see, is the youngest yet. If this pattern is true—if it repeated all over the kingdom—then to the south and east of the Lias limestone and the red sandstone and the coal, there must always be chalk. And there," he pointed to the eastern horizon again, as if to keep the vision of the hills alive, "as on the downs of Wiltshire and Hampshire, is the chalk, Mr. Palmer."

"Then you are to be congratulated on a notable discovery, Mr. Smith."

The day of that visit to the thousand-year-old minster goes unrecorded, by Palmer, by Perkins, and by the historians of the great church. Smith himself makes the briefest mention—"from the top of York Minster I could see that the Wolds contained chalk in their contour"—a remark that allows what Roger Osborne suggests to sound more than plausible. It further allows his solid and reliably old-fashioned Mr. Samborne Palmer to write in his supposed diary that for this reason, that unremembered day on the tower of the greatest church in northern England, remains the birthday of the science of stratigraphy. We do not know the precise date; but we do know that the event took place, and that, with Smith at last revealing his theories to two members of the more general public, the science he had for months been on the brink of creating was now announced, and so now existed.

The word *stratigraphy* itself was not to appear in print for another seventy years, and that in a description of Smith's own work. The word *stratigraphical*, however, was to appear a good deal sooner—in 1817, in fact—and it did so as the title of a book by Smith, in which he expatiated on the ideas he first conceived in High Littleton, which possibly—just possibly—he first made public on that cool and windy late summer's day, two hundred feet on top of the great cathedral in the city of York.

✣

And here in the narrative certain fact takes over once again from supposition. The diaries and the memoirs offer fragments from the journey. The trio made a few more northward miles, with Smith now fully confident of his thesis and eagerly and loudly proclaiming his observations and thoughts to all who would listen. His diary interweaves as diaries do the trivial with the profound. He notes how they dined on "pine-apples at the Black Swan," but that when they left town next morning, how geologically similar were the Hambleton Hills to the Cotswolds, three hundred miles back to the southwest; how there were cliffs of the familiar red marl near Thirsk; how there was a yellow limestone lying unconformably above the coal at Ferryhill and

Piercebridge; and yet how in Harrogate, newly opened as a fashionable spa, Dr. Perkins was persuaded to take "a nauseous draft of sulphur-water as we sat in the chaise."

By the end of September the party was back in Bath, and Smith, invigorated and brimming with ideas, settled himself down to the making of the canal. However, matters did not turn out exactly as he had hoped. The engineering work turned out to be very trying, his masters proved exacting, and though by the very nature of his work he was suffused with the geology of the region, he found he had no time to think, to assemble the broader picture. What he had hoped would be an intellectually stimulating time he found to be frustrating. He was obliged, he wrote, to suspend temporarily "my much wished-for opportunity . . . to make an accurate delineation of the stratification throughout England."

However, he was able to mitigate what he thought of as his stratigraphical impotence with the pleasures of burgeoning economic success. Within months of coming back from his tour, he left his rooms at Rugborne Farm and took a lease on the central house in a small and elegant Georgian crescent on the hills just outside and overlooking the city of Bath itself. He was now being paid well—a guinea a day plus traveling expenses—and he had few outgoings and no family to support. He was able to dress as befitted a gentleman of the time, and to engage in a modest round of social events—though so far as can be gathered from his diaries, he tended to restrict himself to the gentleman-scientists of the town, the divines and the antiquaries and the men of odd enthusiasms.

He had no eye for art, no ear for music, he was not socially confident enough to venture to the salons and the tearooms, and he resisted all temptation to become a fop, a dandy, or a dilettante. He was an engineer, a surveyor, a man of minerals and, as he saw it, of great scientific ideas. He was content simply to work with the canal excavators, who called at his house each day in response to advertisements he had placed in the *Bath Chronicle*, and to travel with them to superintend the painstaking, inch-by-inch digging of the works; and he was then happy to come home and gaze from his windows at a prospect that clearly pleased:

> From this point the eye roved anxiously over the interesting expanse which extended before me to the Sugarloaf Mountain in Monmouthshire, and embraced all the vicinities of Bath and Bristol; then did a thousand thoughts occur to me respecting the geology of that and the adjacent districts continually under my eye.

The city of Bath is of course very proud to have had William Smith as a resident, and in 1926 it unveiled a plaque to him. Local worthies explained why the city had so eminently suited him. "Of all the countries with which I am acquainted, no one is so interesting to the geologist as the vicinity of Bath, because in no other are so many strata exposed to view," said one. And another:

> I need not elaborate the physical circumstances which favour the student of geology in Bath; besides the water supply, hot and cold, the steep cliffs and hillsides with their quarries of building stone, the neighbouring coal measures and the canals. Within easy riding distance are outcrops of the stratified rocks from the Silurian to the Upper Cretaceous, those to the east displaying a regular and obvious succession, those to the west disturbed by unconformities and faults.

Stimulating and fashionable though the city might be, Smith evidently chafed—either at the daily commute by horseback to the digging sites, or because of a more general longing for the rural life to which he was so accustomed. With a year of leasing the house in Cottage Crescent* he had taken semipermanent

*The row of seven houses still stands, renamed "Bloomfield Crescent." It is a perfect specimen of Bath Georgian architecture, but it now stands isolated, an oasis of old civility among a great wen of the new and the semidetached.

rooms in the old Swan Inn at Dunkerton, which is right on the route of the canal (and also where the Roman road, the Fosse Way, crosses the canal's route; and where the new cuttings revealed a Jurassic passage, the transition from the Lias in the west of the village, to the inferior oolite, and the limestones of which all Bath is made).

And then, eighteen months or so later still, the restless Mr. Smith moved yet again. For sixteen hundred pounds—three hundred paid as cash, the rest borrowed on a mortgage that was to have the direst of consequences for him many years down the line—he bought his first-ever property, a small and exceptionally pretty estate known as the Tucking Mill.

There has been some confusion as to exactly which house it was he bought. The two-story Tucking Mill Cottage, with its unusual Gothic sash windows, that stands on the narrow, leafy

Tucking Mill—currently wrongly identified (by a plaque, still visible at the bottom right hand of the house) as Smith's home.

Tucking Mill House, nearby, was in fact where William Smith lived.

road between Midford and Monkton Combe—and that has a memorial tablet in its front wall saying that Smith once lived there—seems not to be it. Instead, shrouded by trees to the east of this house is a much plainer and more severe structure, the Tucking Mill House. A tiny brook separates the two properties and, as it happens, this same brook divides two parishes—the one to the west where the tableted cottage stands is the parish of South Stoke, while that with the unmarked, plainer house is the parish of Monkton Combe. And a search of the tax returns shows that it was property in the Monkton Combe parish, owned by a Mr. E. Candler, that Smith bought in March 1798. He must, therefore, have owned the house. A little local controversy still simmers, suggesting that the plaque be moved, in satisfaction of historical accuracy, if not necessarily in the interests of local real estate prices.

In any event it could hardly be more appropriate than for Smith to live in a Tucking Mill. *Tucking* is the old word for "fulling," which is the process whereby wool is scoured, beaten, and cleansed of the lanolin grease with which sheep make themselves warm and waterproof. The substance used to wash the raw wool is to be found, uniquely, in the very Jurassic strata that Smith was slicing through with his canal—a clay, found in the

Middle Jurassic between the inferior and the great oolites, and known as fuller's earth. It is a strange claylike rock, rich in a hydrous aluminum silicate mineral known as smectite, which happens to have the ability to absorb oil. To live in a house that is named after a process in which geology plays an essential part brings a fine symmetry to William Smith's living condition, and one that almost certainly contributed to his eagerness to buy the house in the first place.

Whether it was a wise and prudent decision he would not know for more than twenty years. Yet, as the eighteenth century was coming to a close, Smith had at least all the outward trappings of gentlemanly achievement. He had an excellent job, a group of admiring and influential friends, and now at last he had an elegant house with a small lake, a waterfall, and seventeen acres of well-laid-out grounds.

He was able to use it as his headquarters—not merely for his work but to house the collection that was now fast becoming central to all his activities, and his employment of which is central to the modern memory of him. Tucking Mill House is where William Smith set aside a room, and had carpenters build for him glass-fronted cabinets, so that he could house, collate, catalog, organize, and display his growing collection, from the rocks around him locally, and from his travels far and wide, of fossils.

The unique arrangements of fossils, as he had first realized as he emerged from the collieries at High Littleton, were what enabled him to tell one stratum from another. It was the arrangements of fossils that would empower him to predict what was underground where, and to make a map of it all. In all his searches to come, fossils would be the key.

8

Notes from the Swan

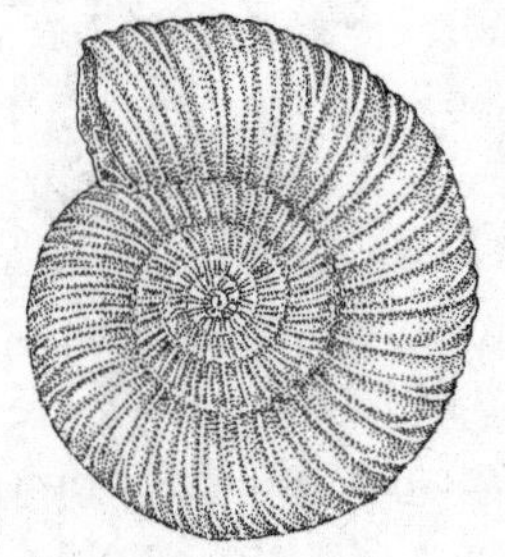

Parkinsonia parkinsoni

In eighteenth-century Britain it was a mark of refinement and impeccable good taste to own and display a collection of fossils. Not only were the objects themselves rare and beautiful, well worthy of display in specially constructed glass cabinets; the simple possession of them hinted at a thirst for knowledge, an awareness of natural philosophy, a sympathetic understanding of the mysterious processes of the earth. And gradually it was from within the world of fossil hunting—a world that would soon be inhabited most prominently by William Smith—that the ideas emerged that would eventually lead Charles Darwin and Alfred Wallace to reach their profound conclusions about the origins of species.

Perhaps for the British *boulevardier* in the eighteenth century, the interest in fossils was for their beauty and rarity, little more. The items, be they small or large, plant or animal, or merely the mysterious results of the fossil-making "plastic force," would be displayed with reverence, handled with delicacy, viewed with awe. Collectors of fine jade today are a fair comparison with those of fossils two centuries ago—in that they are proud and protective, given to learning and (usually) the possession of some social standing. The clear and important difference is that the intricacies of objects made of jade are the artifice of human beings, while the strangely beautiful shapes and marks that delineate a fossil are the evidence—if ever in eighteenth-century Britain there was agreement on this matter—of the work of God.

The *Dictionary of National Biography* records the occurrence of the plural word *fossils* 293 times, and 177 prominent men and women from British history are listed as having had an interest in, or more likely a collection of, such treasures. Most of the listed collectors appear to have lived between the mid-eighteenth and mid-nineteenth centuries. Few people whose lives are otherwise worthy of recording seem to have collected fossils before 1700; and as with postage stamps and coins, few contemporary amateur fossilists will admit to a mania for collecting them.

Indeed the fashion—for that is all it was, a fashion—began to die in mid-Victorian times. The spread of travel and a growing amazement with the outside world suddenly began to make anthropological souvenirs more valued as icons than dirt-encrusted items from earth history. All of a sudden drawing rooms became places to record and show off the material rewards of journeying through space, rather than the dusty and mysterious objects that came from journeying through time. What had hitherto been a signifier of drawing-room decorum seemed overnight to become the pastime of the dull, and then steadily to evolve into what amateur paleontology is now: no more than the mark of the nerd.

There is much to learn from the *DNB* about the nature and the habits of onetime fossil collectors. The 177 entries show the typical collector of the time to have had certain outward similarities of background, knowledge, and social standing. Most of them—this being the less sexually enlightened end of the nineteenth century—happened to be men, although by chance it was a young Dorset woman who was perhaps the most famous fossil collector of them all.

Mary Anning was thirty years younger than William Smith, and there is no record that the pair ever met—but her birthplace and scene of all her paleontological triumphs, the small seaside town of Lyme Regis, evidently interested Smith: In one of his notebooks there is a rough sketch-map of the Lower Jurassic sea-cliffs there, dated 1794—five years before Mary Anning was born.

Her life was short indeed, even by the standards of the day—and yet the fact that she survived a lightning strike (which killed three adults) when she was a year old always lent locals a suspicion that hers would be an eccentric and furious one. Most of it she spent carefully prying choice specimens of fossil creatures from the Lias cliffs near her home. Her father had taught her something of fossil gathering, since his own business was making the very cabinets in which the well-heeled local collectors would keep their specimens. Her best-known find is the original ichthyosaur, a massive confection of shiny brown bones she first disposed of to the duke of Buckingham, which is now carefully reconstructed in London's Natural History Museum. She was only twelve when she found it, only twenty-two when she discovered a juvenile specimen of the huge marine reptile later

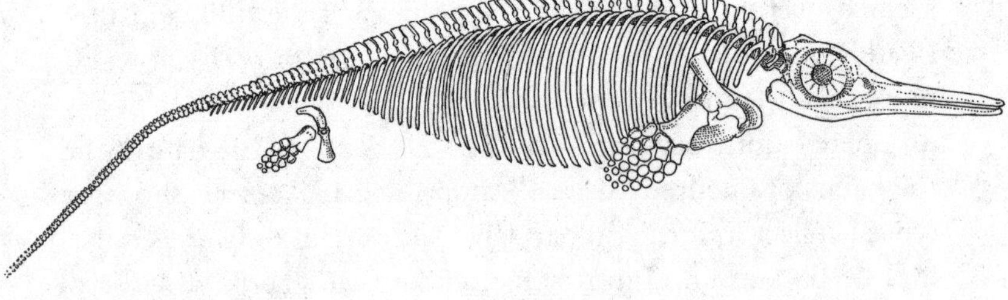

A fossil ichthyosaur.

named a plesiosaur,* and not yet thirty when she found a near-perfect specimen of the bird progenitor, the pterodactyl, and sent it off to Oxford.

For a while this untutored young woman made a sizable income, either by selling fossils to visitors—for whom Lyme Regis is still a major tourist center today—or leading would-be collectors to the cliffs to find specimens for themselves. The names of her customers are like a roll call of the leading geologists—of the day—William Conybeare, Sir Henry de la Beche (who lived nearby), Dean William Buckland. But slowly the popular craze for collecting began to wane, and by 1847, when Mary Anning died at the age of forty-eight of breast cancer, she had been all but forgotten and had passed into obscurity.

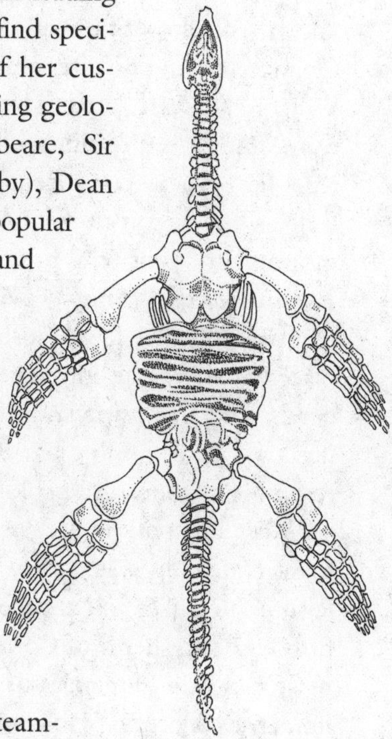

A fossil plesiosaur.

De la Beche, who went on to become the first director of the British Geological Survey, drew a fanciful cartoon for her, showing what Dorset might have looked like in the Middle Jurassic, with enormous and rather genial-seeming monsters rising from the steaming deeps. The drawing became rather popular, and Sir Henry made sure that all the proceeds went to Mary, to help this modest heroine of the science as her fortunes began to decline.

*She wrote to Buckland, the flamboyant and eccentric professor of geology at Oxford, about her discovery of the baby plesiosaur, well knowing that he would find delightful her observation that the animal's neck "had a most graceful curve," and more charming still her discovery that, lodged above its pelvic bone, right where its colon would have passed, was a newly formed coprolite, a fossilised version of the item that, had it lived, the beast was just about to leave steaming in its wake.

There was another woman geologist and collector whose name does not figure in the existing DNB, but should.* She was Etheldred Bennett, a great-granddaughter of a seventeenth-century Archbishop of Canterbury. She was born in 1776, and she definitely met William Smith—indeed, gave him a piece of the well-known Tisbury coral, of which she was England's best-known collector. She made a specialty of exploring the Middle Cretaceous upper greensand in the Vale of Wardour, in Wiltshire: As a relative wrote, "while other ladies of her time were doing needlepoint and chattering over their cups of India tea, she became competent at systematic scientific research, as well as the vigorous fieldwork of fossil hunting." She had a monograph privately printed: *A Catalogue of the Organic Remains of the County of Wilts.* All evidence suggests she died a maiden aunt; her family insisted that one of the specimens later placed in her collection, nestled among her sponges and her corals, and thanks presumably to a cooperative undertaker, was her own heart, unbroken but quite petrified—transformed to resemble a stone, as a geologist's heart perhaps deserves to be.

Most amateur collectors were comfortably established, for fossil collecting was widely seen as a fashion for gentlemen of leisure. Men like, for example, the redoubtable Sir John St. Aubyn, fifth baronet, sheriff of Cornwall until his death in 1839, a grand master of the Freemasons, and a man who augmented his immense collection of minerals by buying for one hundred pounds the entire fossil collection of the remarkable Richard Greene of Lichfield. Greene, so far as we know, was a like-minded swell who had amassed (to the approval of his friend and relative Samuel Johnson) a houseful of "coins, crucifixes, watches, minerals, orreries, deeds and manuscripts, missals, muskets, and specimens of armour," as well as hundreds of ancient shells, graptolite etchings, and ammonites made of iron pyrite.

---

*And indeed will appear in the *New DNB*, thanks to the efforts of her champion, Hugh Torrens.

Then there was, at almost exactly the same time, the East India Company's naval officer, London banker and magnificently named Searles Valentine Wood the Elder, whose curiosity was first stirred while he was convalescing in Norfolk, but who, once recovered, embarked on a lifelong study of the fossil mollusks to be found in the construction sites of London. He was a member of the little-known body the London Clay Club, and wrote book after book on his enormous collection of fossil bivalves, which he eventually donated to the British Museum. The Natural History Museum in South Kensington, where they rest today, is replete with the evidence of a century's worth of enthusiasms like Valentine's—collection after collection, testimony to the value of the amateur scientists who so flourished in this remarkable time in British history.

Many of the most assiduous fossilists were what used to be called "divines"—a curious happenstance, considering the assault that any intelligent understanding of fossils would later have on divinity's most firmly held notions, like the Creation and the Flood. The Reverend Thomas Lewis of Ross-on-Wye is characteristic of the type: He is proud enough to offer a self-description—"geologist and antiquary"—rather than to note his formal position as vicar of Bridstow. His name may be forgotten by the curacy, but it is remembered in at least three Silurian fossil species that were named after him, all of them appropriately worthy (as may befit a clergyman) and rather dull.

Many of the priestly collectors found in fossil hunting a much-needed intellectual stimulus, a relief from the unengaging topics that normally fill a parson's life. The Reverend George Young, from the Scottish village of Coxiedean, was a theologian attracted to the mysteries of fossils. He had been taught by John Playfair, one of the giants of early academic geology, and he came to prominence in 1819 with his discovery, in Yorkshire, of a gigantic reptile ichthyosaur since identified as *Leptopterygius acutirostris.*

Though the find brought the enthusiastic Presbyterian minister some national fame—for a while he was held in almost the same esteem as Mary Anning—it equally confronted him with an interesting challenge, an acute mental and moral dilemma. It forced him to ponder two possibilities that his religious beliefs sternly discountenanced: animal extinction on the one hand (there were no living ichthyosaurs—and so this particular species must have vanished), and animal evolution on the other (the crocodiles and dolphins to which this beast appeared to have been related were much less primitive than this—and so some advances must have taken place over time; the less fit and able must have been weeded out and left behind to die). Consideration of either of these possibilities was a heresy and an anathema to contemporary followers of the Bible, who regarded the great book (as do fundamentalists today) as nothing less than a documentary history of the planet.

The Reverend Young was forced in consequence to engage in some interesting spiritual gymnastics to come to terms with the problem. He eventually committed his conclusions to paper in 1840 in a book with what might be considered the somewhat contradictory title *Scriptural Geology*. The science he advanced in it was not overendowed with logic: The ichthyosaur he had found was not extinct, he declared, because a living specimen would probably be found sooner or later: ". . . when the seas and large rivers of our globe shall have been more fully explored, many animals may be brought to knowledge of the naturalist, which at present are known only in the state of fossils." (It would have amused Mr. Young greatly had he been alive at Christmas 1938, when the first coelacanth was found on the deck of a trawler newly come ashore in South Africa. He would doubtless have thought this vindicated his otherwise dreamily unscientific view.)

And as for evolution—Darwin's theory was not to be outlined for another twenty years, but men like Young, students of the realities of the fossil world, were already moving hesitantly toward the brink:

> Some have alleged . . . that in tracing the beds upwards we discern among the inclosed bodies a gradual progress from the more rude and simple creatures, to the more perfect and completely organised; as if the Creator's skill had improved by practice. But for this strange idea there is no foundation: creatures of the most perfect organization occur in the lower beds as well as the higher.

The Reverend Young could not, however, go any further than this: The forces ranged against him—of custom, history, doctrine, and common acceptance—were just far too formidable.

My own favorite, though sadly no more than a peripheral player in this story, is Samuel Woodward, a Norfolk collector and almost exact contemporary of Smith's who worked for all of his forty-eight years in either an insurance office or a bank. He was fascinated by fossils and built up a large collection. He was not nobly born, however, nor could he have been described as a gentleman for whom paleontology was merely an idle pursuit for impressing the neighbors. He was ordinariness personified: His father had been a bombazine weaver, and his own apprenticeship was in the making of camlets.* Smith would probably have liked him: Both were men of modest beginnings, for whom fossils were more a passion, less a pastime for the *au courant.*

Yet it was not to be one of these modest men but a number of the more gently born collectors and spiritual figures whose influence was eventually to help place William Smith firmly on the flood tide of history. There was William Cunnington, a man still remembered around Devizes as being the antiquary who excavated most of the ancient long barrows with which the chalk downs of the country are littered. It was Cunnington who introduced Smith to the aforementioned Miss Bennett, who fascinated him

*Bombazine is a thick fabric that, in black, is often used as mourning dress; camlets are fine cloths woven from angora or mohair, as fashionable in the eighteenth century as pashmina was to become in more recent times.

with her collection of sponges and corals. The man who would later become the Father of English Geology thus briefly encountered the person who, in some circles at least, is thought of as English Geology's First Woman if not quite (since she remained unmarried, and was described as "somewhat mannish") its mother.

There was the Reverend Richard Warner, a great man for both writing and walking,* but a figure who suffered "severe and reiterated disappointments"—for one of his books was judged a plagiary, another set of volumes was burned by mistake at the printer's, and someone "dressed up as a gentleman" (or so wrote William Smith) made off with his immense fossil collection by giving him a check that then promptly bounced.

There was the somewhat happier Reverend Benjamin Richardson, the rector of the Somerset parish of Farleigh Hungerford. There was also Richardson's longtime friend, the Reverend Joseph Townsend, who was by calling a doctor and a Calvinist minister, then living in Bath, who had been well and expensively educated at Cambridge and Edinburgh. Townsend had traveled widely in Spain, and had brought back hundreds of fossils from the local limestones. He had not, Smith was later to write with relief, drawn any conclusions from his finds, and he was to remark later, and ruefully, "Ah, Smith, were I now to go over to Spain again I should give a very different account of the country."

He would do so because, for the first time, William Smith was beginning to take a keenly intelligent interest in not just the rocks in the cuttings of the coal canal, but of the fossils too. And once he had begun to do so, then who better with whom to discuss his discoveries than the local worthies who had amassed collections themselves? His newfound social standing, his now-close friendship with the widowed Lady Jones of Rugborne, his rela-

*His *Walks through Wales* was an eighteenth-century bestseller; his *English Diatesseron* somewhat less so.

tively good financial condition, his ownership (even if mortgaged) of a small and pretty estate at Tucking Mill, his brief occupancy of a substantial terraced house in Bath itself—all these features commended the uneducated Smith to learned men like Townsend and Richardson, Cunnington and Warner, and allowed them to play a role in his life that he would later acknowledge as of huge importance.

✢

His work on the canal bed and its continuing line of progress was sometimes more confusing than it should be. Smith had no problem recognizing the differences between most of the strata, true: there was a very obvious difference between the red marl and the coal measures, an equally obvious difference between the spawnlike granules in the limestones of the inferior oolite and the arenaceous beds of the Lias. Yet some of the strata through which the excavators were making progress, particularly the finer-grained sandstones, looked too similar. From time to time it proved very difficult, Smith found, to differentiate one bed from another: In one cutting there may have been a sandstone and in another, half a mile away, there may have been another that looked identical—and yet, to judge from a dip and strike that did not vary between the two outcrops, logic suggested that the two formations were not the same at all, had been laid down at different times, and were in fact separated by hundreds, perhaps even thousands, of feet of vertical distance.

To understand the nature of this problem it is perhaps easier to imagine something of the circumstances when the rocks were being laid down. Think, for example, of the conditions in the Lower and Middle Jurassic in North Somerset—something, it is worth remembering, that Smith would have been quite unable to imagine since he had no idea of the ages of the rocks he examined, of the paleogeography of the region, of any of the concepts that permeate modern geology.

He would not have known what modern science allows us to

know, which is that for most of the 51-million-year period of time that began 208 million years ago, when the Jurassic opens, most of North Somerset was covered by a shallow sea, at the western edge of a vast ocean called the Tethys. In addition, since all England was then positioned about thirty-five degrees north of the equator, the waters were subtropical, and warm.

But the sea in those days, much like the sea today, was not uniformly deep, and, since it was at the edge of the Tethyan Ocean, it was at times close to landmasses from which, in places, rivers cascaded or seeped, estuaries were formed, volcanoes erupted, cliffs collapsed, and where currents of sand and water swept down through deep ocean canyons. Paleogeography is a study that involves the constant remembrance of time and space, as well as all the physical conditions in which a particular rock type may be laid down—meaning that at any one time, several

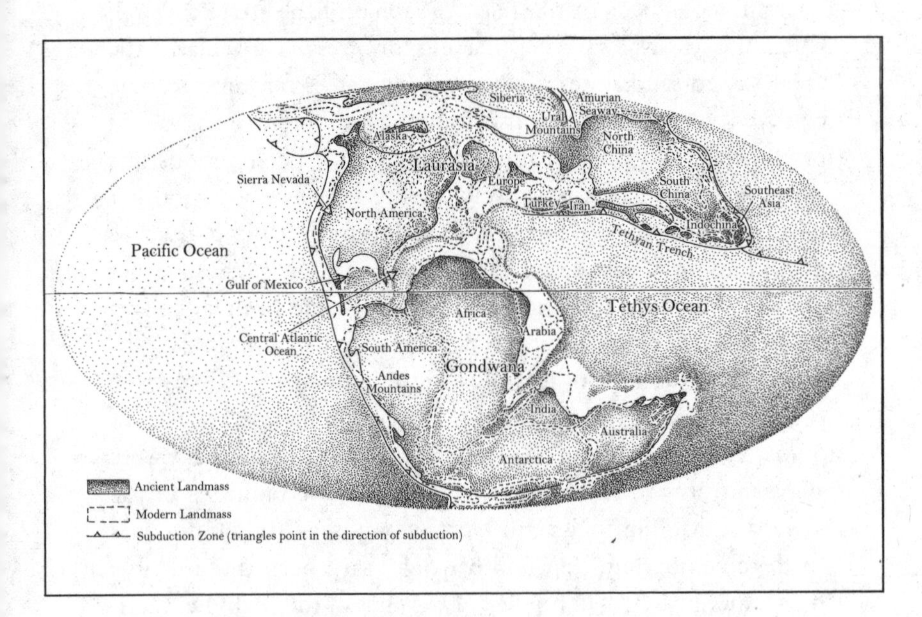

The extent of the Tethyan Ocean in Middle Jurassic times, 152 million years ago.

different rock types may be being deposited or created at different places; and that over any extended period of time the very same rock—or at least, rocks with the very same lithology—may be being laid down at different places.

Hence the confusion. When William Smith was looking at the sandy outcrops of the Upper Lias in a few square miles around the village of Midford, say, he might find a succession of sandy beds in one valley, and another succession of sandy beds in another valley, that looked to all intents and purposes the same, but that his knowledge of their dip and strike and distance apart persuaded him were not the same at all—that the bed lying on top was younger than (that is, had been deposited more recently than) the bed that lay below.

The conditions governing the type of rock, the facies, that had been laid down in each of these two valleys had been exactly the same—they had been deposited near the beach of a warm and shallow sea, with maybe some incoming muddy deposits from a nearby river. But their attitude—going back to the bread-and-butter analogy he had come up with back in his High Littleton days—still applied: they could not have been the same bed of rock, and they must have been separated by scores, maybe hundreds of feet—and hundreds of feet meant at the very least, a long period of time. What the outcrops indicated was two different periods of time, when the same conditions for deposit obtained. How, then, to tell the rocks apart?

The answer lay in Smith's sudden realization that there was just one aspect of the two types of rock, and only one, that differed. The blocks of stone found in the cuttings may have all had the same color, an acid bottle would show them all to have the same chemistry, a magnifying lens would show the sandstones as all having the same grain size. But the fossils that were to be found in the two rocks—the bivalves, the ammonites, the gastropods, the corals—*they were all subtly different.*

Every single one of the specimens of one kind of fossil might be the same throughout one bed, but would be subtly different from those of the same kind of fossil found in another bed. A period of time would have elapsed between the deposition of the two beds, and thus a period of time between the existence of the two kinds of animals it embraced. Evolution—we can say this today, but Smith had not even the vaguest conception of it back then—would have occurred. Those animals of which there would be fossilized remains that were found lower down in the series would be more primitive; those found in the rock layers above, less so. But that was not the point. The important discovery that Smith made was that certain beds had certain fossils, that they were unique and peculiar to that bed and to that period of time in geologic history. They were never to be seen again in rocks that came later—in other words, in the rocks that appeared above. They were never seen before, either: They were peculiar, that is to say, to a certain and specific period in geologic time; they were the key to making a positive identification of what one rock might be in relation to any other.

Day after day during the late summer and autumn of 1795, whether he was working surveying the canal or simply clambering over rocks that interested him while his horse champed contentedly beside him, Smith tested and retested his theory. At each outcrop he came to he would gingerly chip and pry and prise as many fossils as he could from their enfolding rock. Each evening he would take his specimens back to his elegant new terraced home in Bath. He would wash and dry each fossil, be it a pedestrian looking oyster shell or the magnificent twirling fantasy of a full-blown ammonite, and lay each carefully, on a pad of cotton, in drawer after drawer of his cabinets, carefully noting the rock, the horizon, the facies, and the lithology from which each came.

And as his systematic collecting proceeded, and as the size and quality of his collection was daily enhanced, so his theory was confirmed and reconfirmed: A layer of rock, it now seemed incontrovertibly true, could be positively and invariably identified simply and solely by the fossils that were to be found within it.

Wherever in the hills around Bath a sandstone appeared with a particular specimen of fossil enclosed within, then it was certain that it was the very same rock, laid down at the very same time. And if this rock-and-fossil assemblage appeared not just in the hills around Bath, but in the valleys of Oxfordshire too, and was found in a quarry in Rutland, beside a road in Lincolnshire, on a peak near York, and finally in a cliff near Whitby, then it, too was the selfsame rock. Not just a similar rock: the same rock. And then, the corollary said, by joining the dots of its occurrence across the land, one could show just where this particular rock occurred all over the nation, and whether it made an outcrop or not. And one could do this not just in the nation, but in theory all over the world. One of the enigmas that was central to an unraveling of the mysteries of the planet had now demonstrably been solved. What he had vaguely imagined might be true when he looked through the mines near High Littleton, was clearly an axiom, a fundamental fact of the new geological knowledge. And he, William Smith, was the first to say so.

Smith was exultant at his realization, and committed his thoughts to paper with excited promptitude. He was in the Swan Inn at Dunkerton, sheltering from the cold on the evening of Tuesday, January 5, 1796. He had decided that evening not to brave the elements, not to go back home to Bath. He took a sheet of paper and wrote in his distinctively bold handwriting a long single sentence. The note survives, its underlining preserved for posterity. It was a sentence that, of all he wrote, is perhaps most deserving to be his epitaph:

> Fossils have long been studied as great curiosities, collected with great pains, treasured with great care and at a great expense, and showed and admired with as much pleasure as a

child's rattle or a hobby-horse is shown and admired by himself and his playfellows, because it is pretty; and this has been done by thousands who have never paid the least regard to that wonderful order and regularity with which Nature has disposed of these singular productions, and assigned to each class its particular stratum.

Later, in more reflective mood, he would write:

For six years I put my notions of stratification to the test of excavation; and I generally pointed out to contractors and others, who came to undertake the work, what the various parts of the canal would be dug through. But the great similarity of the rocks of the Oolite, on and near the end of the canal towards Bath, required more than superficial observation to determine whether these hills were not composed of one, two or even three of these rocks, as by the distinctions of some parts seemed to appear. These doubts were at length removed by more particular attention to the site of the organic fossils which I had long collected. This discovery of the mode of identifying the strata by the organised fossils respectively imbedded therein led to the most important distinctions.

In reflective mood Smith seems more the engineer, less the romantic. In middle age he is, and understandably, no longer quite so astonished at the "wonderful order" that he had realized the fossils displayed—an astonishment of discovery which today remains the most haunting aspect of that hastily scribbled note made at the Swan Inn. But the message remains the same, however eloquent or sentimental the prose. A puzzle had been solved. A riddle unscrambled. Now was the time to make something of the answer.

9

The Dictator in the Drawing Room

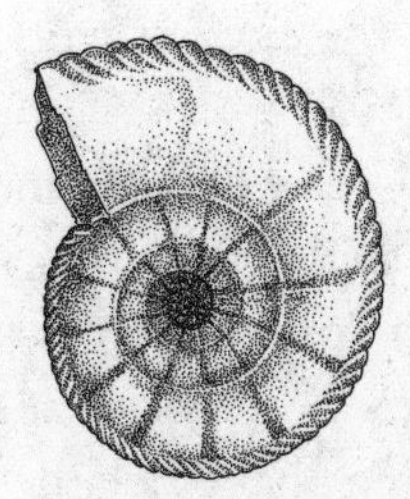

Zigzagiceras zigzag

Some Romans had called what we call Bath *Aquae Calidae*—the hot waters. Others preferred *Aquae Sulis*, naming the scalding springs in homage to the presiding Celtic water deity, later twinned with their very own Minerva. And for the two thousand years since centurions first erected their stone bathing stalls at this most convenient stopping place on the frontier between the Roman and the Celtic worlds, Bath has been an important, memorably unusual, and often very fashionable place. "Oh!" exclaims Catherine Morland in Jane Austen's *Northanger Abbey*, "who can ever be tired of Bath?"

Nowadays it is the combination of social style, elegance, and fine building stone—tales of the costume parties of Beau Nash, the buildings of a famous father and son, both called John Wood, and the honey-colored oolites of the Middle Jurassic—that still, in the main, impress. The tour groups line up in endless succession before the Royal Crescent and the Circus, the Assembly Rooms and the baths themselves, eager to revel in the pleasures of majestic architecture and public grace. Tens of thousands of visitors throng the streets, passing briefly through the compact little city, within its amphitheater of hills. It is, for tourists, one of the score of essential English way stations between the great pile of Buckingham Palace and the artless country cottage of Anne Hathaway.

A few visitors come to stay, and some to study, and an even smaller number to take the hydropathic cure by drinking some of the most foul-tasting mineral waters with which the mantle's heat has ever supplied us. "Particklery unpleasant," Sam Weller had said a century ago, in *The Pickwick Papers.* "A wery strong flavour o' warm flatirons." But efficacious, they used to say in the century before Dickens, and in those days the finer folk of England would flock to Bath in their thousands, and lodge, imbibe, and amuse themselves, and amuse all the envious world that looked on.

But people also came to Bath to study, discuss, debate, and argue. The citizens of Bath—the population in 1800 had risen dramatically, to thirty thousand—liked to think of themselves as inhabiting the nation's second city, in matters both social and intellectual. Few of the citizenry had forgotten that Adelard—"England's first scientist"—a twelfth-century philosopher who had written treatises on the abacus and the astrolabe, had been born in Bath.

Only London attracted finer minds, just as only London had grander parties and soirées. And so in 1777 a move was made to establish and formalize the intellectual ambitions of the city, by creating a society that had as its sole purpose the encouragement of the discussion and dissemination of ideas. The *Bath Chronicle* of August 28 carried an advertisement, placed by a weaver's son named Edmund Rack, and directed at "The Nobility and Gentry in the Counties of Somerset, Gloucestershire, Wiltshire and Dorset in General, and the Cities of Bath and Bristol in Particular."

They were to consider, Rack wrote, a proposal for "the institution of a Society in this city, for the encouragement of Agriculture, Planting, Manufactures, Commerce and Fine Arts." The tone, flattering and seductive to the city's elite, evidently worked: Twenty-two of the noblest and gentlest-born of Bath's citizenry turned up at a meeting held a week later in what would later be the Royal York Hotel, and the first of the distinguished intellectual societies for which the city would become famous was formally constituted.

An impressive roll call of luminaries chose over the years to become associated with or full members of the various new bodies—the Bath and West of England Society, the Bath Agricultural Society, the Bath Philosophical Society, the Literary Society, and today's successors to them all, the Bath Royal Literary and Scientific Institution and the Royal Bath and West of England Society (now based in Shepton Mallet). There was Joseph Priestley (who discovered oxygen); Thomas Malthus (the economist and population expert), Sir William Herschel (who discovered Uranus* lurking way at the back of the solar system), Humphry Davy (who discovered sodium and potassium), and one Augustus Voelcker, a German, who was a specialist in the chemistry of cheese and set up a school to teach cheesemaking in Wells, nearby.

And on December 22, 1796, it was announced at the annual meeting that, elected unanimously in consequence of his growing reputation for canal making, his expertise in farming, and his keen new interest in his unromantic freelance business of solving problems with the drainage of fields, membership of the Bath and West of England Society was gained by one of the least noble and least gently born men in the city, the blacksmith's son from Oxfordshire, William Smith.

His social standing was improving fast. The same annual

*He wanted to name the first telescopically discovered planet *Georgium Sidus*, in honor of George III, but was eventually persuaded to name it after the old Greek god who had fathered Saturn and the Titans.

meeting recorded that the duke of Bedford, the earl of Egremont, and the earl of Peterborough—farmer-aristocrats all—were in the same company as Smith, and he himself noted that in the years following he came to know each of them well. At the time of his election he had a house on a good terrace in Bath, and his more or less permanent lodgings in the Swan Inn at Dunkerton, which he used when he was delayed in the countryside on canal business. He was well paid, well regarded, sought after. And now, through the Bath Society, he found he was winning friends in influential places—members like the vicars Benjamin Richardson and Joseph Townsend, both of whom were fossil collectors, owners of immense houses in the city—and travelers in the circle of the fashionable; and fellow members like John Billingsley and Thomas Davis—the latter the land steward to the marquess of Bath at his nearby estate at Longleat—who were at the time engaged in writing exhaustive studies of the state of local farming for the Somerset and Wiltshire Boards of Agriculture.

It was this latter pair who first introduced William Smith to the notion of making maps. Although, as we shall see, history has been more generous in its assessment of the importance in William Smith's extraordinary story of the Reverends Richardson and Townsend, it was actually Billingsley and Davis who gave him the idea that would be central to his coming achievement. For while Smith had no difficulty at all in displaying the vertical extent of the geology he found—he just drew cross-sections and tables as everyone else did—he had the very greatest difficulty in working out how to display the way in which these strata were exposed horizontally, how the outcrops of different kinds of rock were displayed geographically.

Except that one day in late 1798 he suddenly saw just how he could do it. He was reading the latest edition of the Somerset *County Agricultural Report*, and there, buried without comment among the statistics on pigs and the effects of new cattle crossbreeding programs, Smith found an intriguing small map.

Billingsley and Davis, it turned out, had sketched their latest in a series of maps for the report that showed, crudely but effectively, the geographical extent of each of the various soils and types of vegetation that were known in the countryside around Bath.

The maps were detailed and, moreover, they were *colored*: with the use of blues and yellows and greens, all painstakingly applied by hand, they showed the local forests, meadows, pastures. The maps displayed graphically all the nearby hills, rivers, and lakes. And, most important for Smith, they hinted at what lay underneath the surface of the earth, by showing, also in colors, the outcrops of the red earth, the courses of the coal.

In a flash Smith now realized the possibilities. If ordinary agricultural men like Billingsley and Davis were capable of making maps that could display such details, then, with his even greater graphical skills and now a good deal of new and detailed knowledge about just what lay beneath the surface, he himself could draw charts that would show the courses followed by all the rocks that he knew lay down below.

He could use his skills and unusual knowledge, in other words, to draw a brand-new map the likes of which had never been known. He could draw a chart of what could not be seen. And in doing so he could create what had never been created before—*a true geological map.*

His diaries and notes showed that he then puzzled over the finest details—most important, whether he could make the maps relatively inexpensively, by drawing the outcrops in black and white, by using lines of different thickness or by using cross-hatching, to illustrate the different rocks. But he decided he could not. Color, costly though it was to print, and time consuming to apply, was in his view essential for a chart that would be so complex as a map of the unseen underworld. He thus embarked upon a technique of coloring that he was to embrace for the following thirty years of his cartographic career.

He decided first to start his mapping by applying his new

techniques to what he knew—the area in the immediate vicinity of Bath itself. By happy chance in the early summer of 1799 a new book was published, *The Historic and Local New Bath Guide*, printed by A. Taylor and W. Nayler, Booksellers. Its frontispiece turned out to be a handsome map of the city—a map Smith immediately felt he could use as a base on which to superimpose what he now knew about the geology.

Taylor and Nayler's map was somewhat unusual in appear-

William Smith's original circular map of the geology around Bath, published in 1799, and thus technically the oldest of all true geological maps.

ance, not least because it was circular, about fifteen inches in diameter. It was on a scale of one and one-half inches to the mile. Bath lay in the center, like a bull's-eye. The Avon wandered across from northwest to southeast. The Kennet and Avon Canal was marked, as was the still-not-quite-completed Somerset Coal Canal. The countryside for five miles in either direction was depicted in some detail, with grand houses, stands of trees, parish churches, the turnpikes and common roads, and "with Alterations and Improvements to the present Time." It was uncolored and, despite holding plenty of important information was designed in a nicely uncluttered way. For William Smith's purposes it was ideal.

He promptly set to work in his Cottage Crescent house, transferring all the notes from his survey books—oolite with this particular ammonite here, Lias with this *Lingula* there, red marl with these *Ostrea* in this valley, river deposits with clamshells here—onto the base map itself. He extrapolated his dip and strike details, made some logical postulations about where the various strata might end up, then joined the dots—and found he had created on the map a number of shapes, all enclosed and irregular-shaped bodies. They were bodies of which he could now say, and with certainty—*this* one shows where the oolite exists, *this* is the location underground of the Lias. It had never been done before: The unseen world of the underground was all of a sudden on display, seeable, meant to be seen, the hitherto invisible made visible at last.

And to make it not just visible but startlingly apparent to anyone who glanced at his map, Smith then mimicked the technique of Billingsley and Davis and hand-colored the different bodies he had drawn. He colored the outcrop of the oolite a rich shade of yellow; the Lias the dirty blue of one of its building stones; and the red marls of the Trias a brickish red. It was a color scheme that, as it happens, has remained in place in almost all geological maps to this day.

By the middle of the summer of 1799, all was done. What resulted from William Smith's labors was a map that, for all its age and weatherbeaten look, still has a strangely ethereal beauty. It may not have been of very great use: It was very limited in extent, it showed the outcrop of only three types of rock, and since it had no index it was hardly much of a guide to the underside of the Bath countryside. But the map hangs to this day in what are still called the "apartments" of the Geological Society of London in Piccadilly, and though it is dwarfed by its more famous successors and therefore rarely noticed, it amply deserves to be memorialized. For it is arguably the oldest geological map worthy of its name in existence—primitive, local, and small-scale, true, but nonetheless the oldest, the *ur*-map. The rubric is mostly engraved: "A Map of Five Miles round the City of Bath, on a scale of one inch and a half to a mile, from an Actual Survey, including all the new roads, with Alterations and Improvements to the present time, 1799. Printed for and sold by A. Taylor and W. Nayler, Booksellers, Bath." There is a handwritten addition, in the elegant cursive hand that over the coming years would become so familiar: "Presented to the Geological Society, February 18th, 1831. Wm. Smith, Coloured Geologically in 1799."

William Smith was to make still more history during that fateful year. Mary Anning may have been born that year, Vesuvius may have been erupting, the French Revolution may have been ending. But at 29 Great Pulteney Street in Bath, on the cool evening of Tuesday June 11, 1799, history was being made at a small dinner party. There were only three guests—the Reverend Joseph Townsend, the Reverend Benjamin Richardson, and Smith—the "triumvirate," as one historian was later to say, three of the leading players in the heroic age of geology. As the party drew to a close Smith is reported to have stood up, by invitation, and dictated to his host a document that is still regarded as one of the classics in the annals of science.

Townsend, whose house it was and who was thus the persuasive genius behind Smith's decision to commit his ideas to paper, was in all senses a clever, very well-connected, and most unusual man. He had trained as a doctor, but had then taken holy orders and moved in to the great rectory at Pewsey in Wiltshire, which had been bought for him by his father, who was the local member of Parliament. He appears to have had the breadth of intellect for which the era's intellectual aristocracy was renowned—he taught himself Hebrew, wrote a large book on philology, was fascinated by canals (which is why Richardson first introduced him to Smith at a Bath Society meeting), and was an active member of the regional Highways Trust, which supervised the local turnpikes.*

Townsend had an electrifying manner as a preacher—"his voice at all times sepulchral, but when exerted, of passing loudness admirably adapted to arouse, denounce and alarm"—and for a while was appointed personal chaplain to the headstrong Methodist aristocrat-evangelist Selina Hastings. But then he had a serious falling-out with her over where and when he could preach, and for several years in the 1750s he and a group of like-minded holy men who had been chaplains at her huge collection of churches were pursued by her agents in a fantastical cat-and-mouse game as they preached, without her permission, in barns and open fields and under market crosses, all across the English West Country. A play was written about the affair, with Townsend satirized as the Reverend Timothy Wildgoose.

He eventually fled to Ireland in 1761, and in County Kerry took up an interest—which one might think odd for a doctor and man of the cloth—in a mining company.† He turned out to be fascinated by minerals and by fossils, and managed for the rest

*He was immensely tall and well built, prompting his Highways Trust colleagues to call him the Colossus of Roads.

†Though perhaps not so odd, since his father, Chauncey Townsend, MP, owned several tin and lead mines in Devon and Cornwall.

of his days to commingle his studies of evangelism with those of geology—to the extent that when in 1790 he published the one book for which he is known, *Journey through Spain*, it was crammed with information about the geology and mineralogy of the Iberian Peninsula. Considering the fine irony involved, it bears repeating that the first remark he made when Richardson introduced him to William Smith was how much better his Spanish book might have been had he known of Smith's theories about the strata, superposition, and the use of fossils in recognizing what and where they were.

Richardson himself was a much less daunting figure—the vicar of Farleigh Hungerford (a living—as the job of vicar in a rural parish of the Church of England was then and still is known—in the patronage of one Joseph Houlton, who was in turn to become a keen supporter of Smith and his mapmaking endeavors), and the owner of a huge old library and collection of fossils and rock samples in his home, Farleigh Castle. Richardson met Smith at a gathering of the Agricultural Society. They talked about fossils, and Richardson invited the younger man home to inspect the collection he had amassed. It was then that the first connection was made, the serendipitous meshing of one set of skills with another.

For while Richardson had a magnificent collection of fossils in his town house on Lisbon Terrace, Bath, he had no idea which belonged where in the order of strata, no clue as to which lived earlier, which later. All the fossils were as a result jumbled together, or put into drawers not according to their relative age but according to their type—all the ammonites placed on one shelf, all belemnites on another, all crinoids here, all graptolites there, all brachiopods in one compartment, and so on.

When Smith inspected the fossil cabinet he was both appalled and challenged. He told Richardson that, if he was allowed, he could organize the vicar's collection so that all the fossils would be arranged in their correct, logical, properly created stratigraphical order—with the oldest and least advanced at the bottom, the most sophisticated and complex and youngest at the top. Richardson, who could barely tell one fossil from another, readily agreed. The young man set to work—and within a day the entire amassment of the amateur paleontologist-vicar had been rendered into a scientifically accurate column by the efforts of this extraordinary surveyor-professional.

Richardson was, by all accounts, astonished. No diary records his words, but Smith later said that he explained to Richardson that what he had shown him applied everywhere, and not just to the fossiliferous rocks and specimens in this one collection. No, he said—and this, for the first time to a man with at least some scientific training—*the same strata are always found in the same order of superposition, and they always contain the same peculiar fossils.*

Richardson grasped in an instant the importance of what was being said. He immediately made contact with his friend Townsend, who suggested that Smith's theories be put to the test. If Smith was right, then he should be able to predict with some accuracy what rocks would appear, and what fossils would be found, on both the slopes and at the summit of a prominent nearby hill where Dundry Church had been built.

Smith rose to the task. From his calculations he reckoned the hills would be capped by the limestone of the inferior oolite; and that on the western side of the hill it would be possible to find a series of very specific fossils, peculiar to what is now known as the Lower Jurassic. The three men hired a horse-drawn carriage and hurried northwestward to Dundry Hill. Every few minutes they stopped the coach, climbed down, inspected the outcrops of rocks—and on every single occasion, in terms of outcrop, thickness, lithology, and fossil content, William Smith was right.

"The effect of this . . . was decisive," wrote a contemporary biographer of Smith.

> In general literature and especially in natural history Mr. Smith was immeasurably surpassed by his friends. But they

acknowledged that from his labours in a different quarter, a new light had begun to manifest itself in the previously dark horizons of geology, and they set themselves earnestly to make way for its auspicious influence.

The dinner at Great Pulteney Street that evening was a classic example of such men "earnestly making way" for the coming effects of Smith's thinking. It is not known what was eaten or what if anything was drunk. What is known is that the three men spent much of the time discussing what they had found at Dundry, what Smith had accomplished with the collections at Fairleigh Castle, what theories had now been tested, proved, and confirmed. It remained now, at least in the view of Joseph Townsend, to make a tabular list of the strata of which the three men were well aware—a summary, in fact, of all they knew about the region's rocks, and where each fitted in with respect to all the others.

Richardson and Townsend sat at the dining table, now cleared of all china and glassware. A large piece of paper, the size of a blotting sheet (about twenty-six by thirty-four inches), was placed on the surface. Townsend took out his quill pen and sand shaker. Smith gave him a ruler, and with this he drew five long lines across the page—one horizontal, four vertical. Above the horizontal lines, and in the five boxes thus created by the intersections with the vertical rules and the paper edges, Townsend then wrote, according to Smith's dictation: "Strata, Thickness, Springs, Fossils, Petrifactions &c, &c.," and finally, "Descriptive Characters and Situations." Smith then cleared his throat, and began his dictation.

The list that followed enumerated twenty-three horizons—twenty-three bands of rock that were sufficiently different from one another, and recognizable from the fossils within them, to be counted as separate and unique. The list began with the youngest, and it ended with the oldest. It began with the chalk, which Smith listed as number 1, and ended with the coal, squeezed in almost as an afterthought at the very bottom of the page, as number 23.

In between were twenty-one strata of rocks that had, in many cases, hitherto passed unnamed and unnoted. Now, at a scratch of Joseph Townsend's pen and a shake of blotting sand, they were formally christened by William Smith. Not all were given names—the first four below the chalk were merely described by their lithologies. The names that some older horizons were given did not go down well. The geological establishment of the day suggested that some of the appellations were crude and, worst of all to the refined manners of the day, abominably ugly. Nonetheless, in many cases they have stuck, and at least four of them remain in scientific use in England today.

Below the 300 feet of chalk, Smith declaimed before the others, were first 70 feet of sand. Then 30 feet of clay. Then 30 more feet of clay and stone. And 15 feet of clay. Then 10 feet of the first of named rocks, forest marble. And 60 feet of freestone. A narrow band, no more than 6 feet thick, of blue clay, and then only another 8 of yellow clay. Then, most familiar still today, 6 feet of what would be called fuller's earth. And 80 feet of what Smith called bastard fuller's earth, complete with fossils—"striated cardia, Mytilites, Anomiae, pundibs, and Duck-muscles"—that the narrow band above did not contain.

Below that came the stratum labeled number 12, 30 Feet of Freestone. Then 30 Feet of Sand, 40 of blue marl, 25 of Blue Lias, 15 of White Lias, 15 of Marl Stone, 180 feet of the much-vaunted Red Ground (or Red Earth, or Red Marl, depending on where Smith was writing, and to whom), and "milstone [*sic*]," followed by an unrecorded thickness of Pennant Stone, then a band of Greystones, another of Cliff—whatever that might be (though whatever it was was crammed with "ferns, olive stellate plants, *Threnax parviflora*, or dwarf fern palm of Jamaica)"—which trended, almost imperceptibly, into the coal.

The biggest change between any of the strata was the one that occurred between the Millstone and the Pennant Stone. Beneath the Millstone bed, dictated Smith, "no fossil, shells or animal remains are found; above it, no vegetable impressions." The boundary between the two, Smith had already noticed, was unconformable. That it had so vastly different a fossil population—only animal remains above, only plant remains below—indicates something we know only too well today: that this unconformity marks a vitally important geological boundary.

It was the base of the Permo-Triassic period of geologic time, and the top of the Carboniferous period. It was and is, in other words, a highly observable, universally recognized, and now internationally agreed moment in the earth's long and unstoppable history—and William Smith, Joseph Townsend, and Benjamin Richardson, were the first men to witness the newly discovered fact of this moment being committed to paper. The record written that evening would ensure that this geological junction point would remain in human beings' knowledge of their world, and of themselves, forever.

For the first time the earth had a provable history, a written record that paid no heed or obeisance to religious teaching and dogma, that declared its independence from the kind of faith that is no more than the blind acceptance of absurdity. A science—an elemental, basic science that would in due course allow mankind to exploit the almost limitless treasures of the underworld—had at last broken free from the age-old constraints of doctrine and canonical instruction.

From now on—armed with a new knowledge and understanding of how matters were arranged below the earth's surface—human beings could begin to explore their planet from a different perspective, and with an intellectual freedom that would in time permit them to look for and then to find astonishing things. The reverberations of that late-evening meeting can be felt distinctly down all the years. Each time a new oilfield is opened, or new gold is added to a reserve, or when more platinum or cerium or iron or manganese is won from the earth's crust, it is perhaps appropriate to remember these three men. To remember them, and to savor the irony that, while Townsend and Richardson worked that night under the leadership of William Smith, whose agnosticism was well known, they themselves were churchmen—making this particular bid for intellectual freedom an act of brave defiance, and one of which their bishops would doubtless disapprove. Yet though the Church may have briefly frowned, all humankind went on to benefit.

The trio finished the document at midnight, and made three fair copies. They called it the "Order of the STRATA and their embedded ORGANIC REMAINS, in the vicinity of BATH; examined and proved prior to 1799." Wrote a Bath city historian

THE ORIGINAL TABLE OF STRATA NEAR BATH

The barely legible Table of Strata, dictated by Smith to his two friends in Bath, now preserved at the Geological Society of London.

of the time, "Each person took one copy" and was encouraged to make further copies as necessary. There was, the historian added, "no stipulation as to the use which should be made of it, and accordingly it was extensively distributed, and remained for a long period the type and authority for the descriptions and order of superposition of the strata near Bath."

We are not sure exactly who received copies, except in one or two cases: Benjamin Richardson, for example, acting to fulfill Smith's wishes that the discovery be made universally known, gave "without reserve, a card of the English strata to Baron Rosencrantz, Dr. Muller of Christiania,* and many others, in the year 1801." Smith was pleasantly surprised a couple of years later when a geologist named William Reynolds turned up in Bath from Coalbrookdale in the West Midlands, and told Smith that copies of his table of strata, of which he had one, were circulating among the educated classes in both the East and West Indies!

William Smith had made his observations, had formulated his theory, had tested his ideas, had been proved correct, and was now publishing his notions for the world to read. He was already socially established; now he was on the verge of what no Oxfordshire country boy could ever have imagined: adulation—and fame.

Yet at almost the same moment as he verged on triumph, so William Smith was sowing the wind, of which he would eventually reap the whirlwind. He had, it will be remembered, bought for himself an attractive small estate, the Tucking Mill House, at Midford. He had a mortgage, for the not-insubstantial sum of £1,300, and he was under contract to pay an annual sum to the owner, a Mr. Conolly.† At the time he made the purchase he was

---

*The former name of what was to become the Norwegian capital city of Oslo (as well as, derivatively, an early word for a parallel turn on skis).

†There is some confusion about the precise spelling of the landlord's name: The unusual single *n* spelling appears in all the latter court papers, when the relations between Smith and the owner of Midford Castle had deteriorated.

fully employed by the Somerset Coal Canal Company and had both a generous salary and guaranteed expenses.

But on Wednesday, June 5, 1799, William Smith was abruptly and unceremoniously fired. It has never been entirely clear why. One suggestion has it that he disagreed violently with the way the canal was being built, and in particular over the decision to construct a mechanical lifting device instead of a flight of locks to bypass a hill near the village of Combe Hay. A more popular explanation is that the directors found an unacceptable conflict of interest in his purchase of Tucking Mill—which lay beside the canal—at the very time he was engaged in buying land for the canal's passage. It seems that he approached Conolly with a view to buying land for the canal right of way and asked him if he might throw in a few extra acres for himself at the same time—if true, a somewhat imprudent act that might reasonably be expected to at least raise some canal company eyebrows.

Whatever the reason, the company asked him to leave, quickly, and with little fuss or ceremony and no suggestion of a pension or golden handshake. A job that had brought him almost £450 a year was brought to a sudden end. The firing, even though Smith does not appear to have contested it, unsettled him gravely, and his letters of the time show him to have become briefly angry and embittered. But by the end of the year he seems to have settled somewhat, and by the time he had his famous dinner at Great Pulteney Street, he was his old self, brimming with self-confidence and ambition.

✢

Except that in retrospect there appears to have been more than a touch of hubris about him. For below the surface matters were beginning, if slowly, to unwind. His financial affairs were starting to unravel. Those who would eventually come to cheat him, and try to deprive him of the honors he was rightly due, were beginning to gather, and to circle. He would never have full-time, fixed employment again—his six years with the canal company marked the summit of his career as a company man. From now on he would be in the perilous position of the freelance mineral surveyor, earning as much as his wits and his contacts might bring him.

He did not know it yet, but he was fast standing into danger. And yet it was just at this time, when matters were beginning to go awry, that he began work on the biggest project of them all. He had made the first geological map in the world, of the country around Bath. Now he began to consider the possibilities of making a map, not just of a region, but of an entire country. He wanted nothing less than to know and map the underworld of all of England.

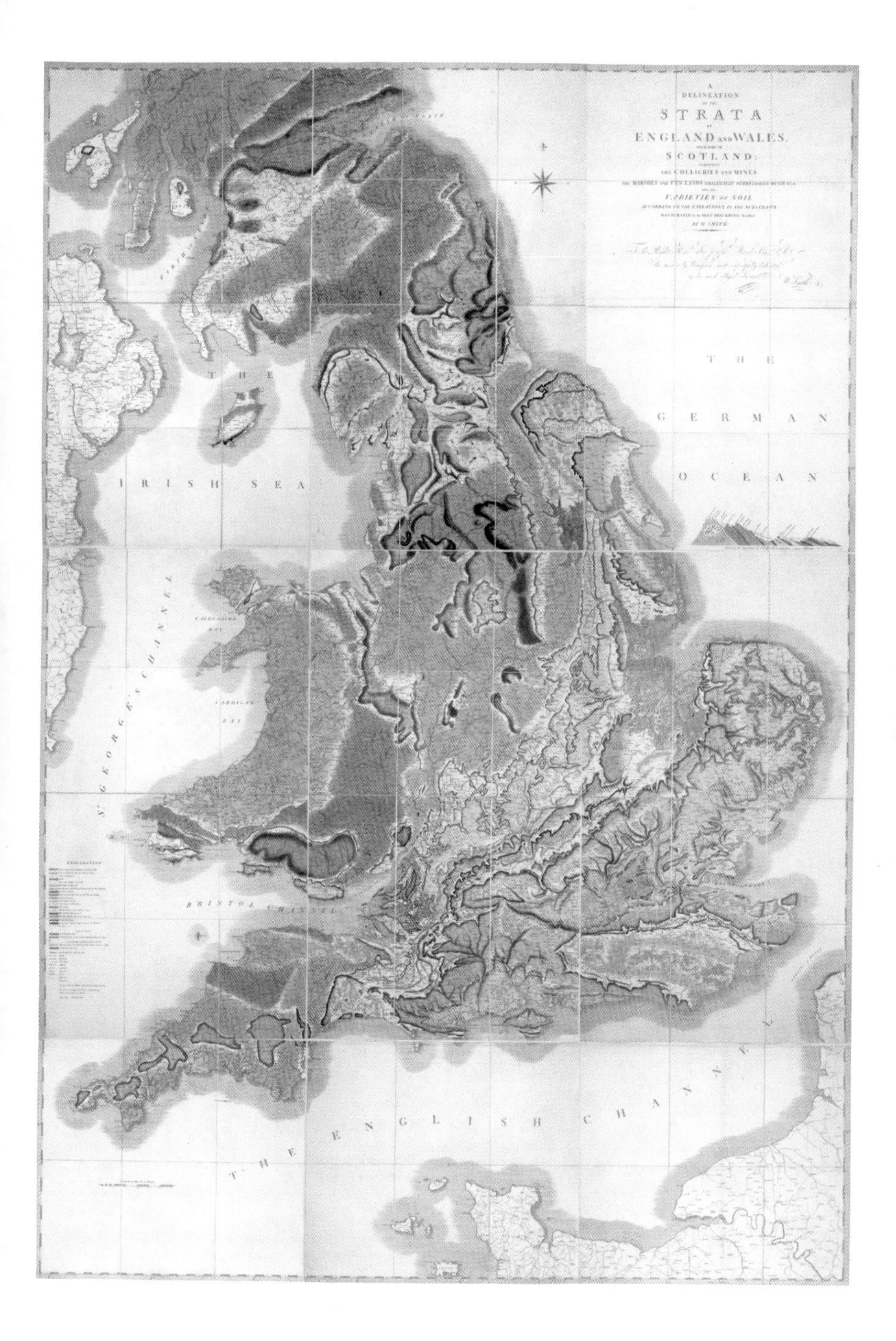
A
DELINEATION
STRATA
ENGLAND AND WALES,
SCOTLAND;
THE COLLIERIES AND MINES
VARIETIES OF SOIL
BY W. SMITH
THE GERMAN OCEAN
THE
IRISH SEA
ST GEORGE'S CHANNEL
CARDIGAN BAY
BRISTOL CHANNEL
THE ENGLISH CHANNEL

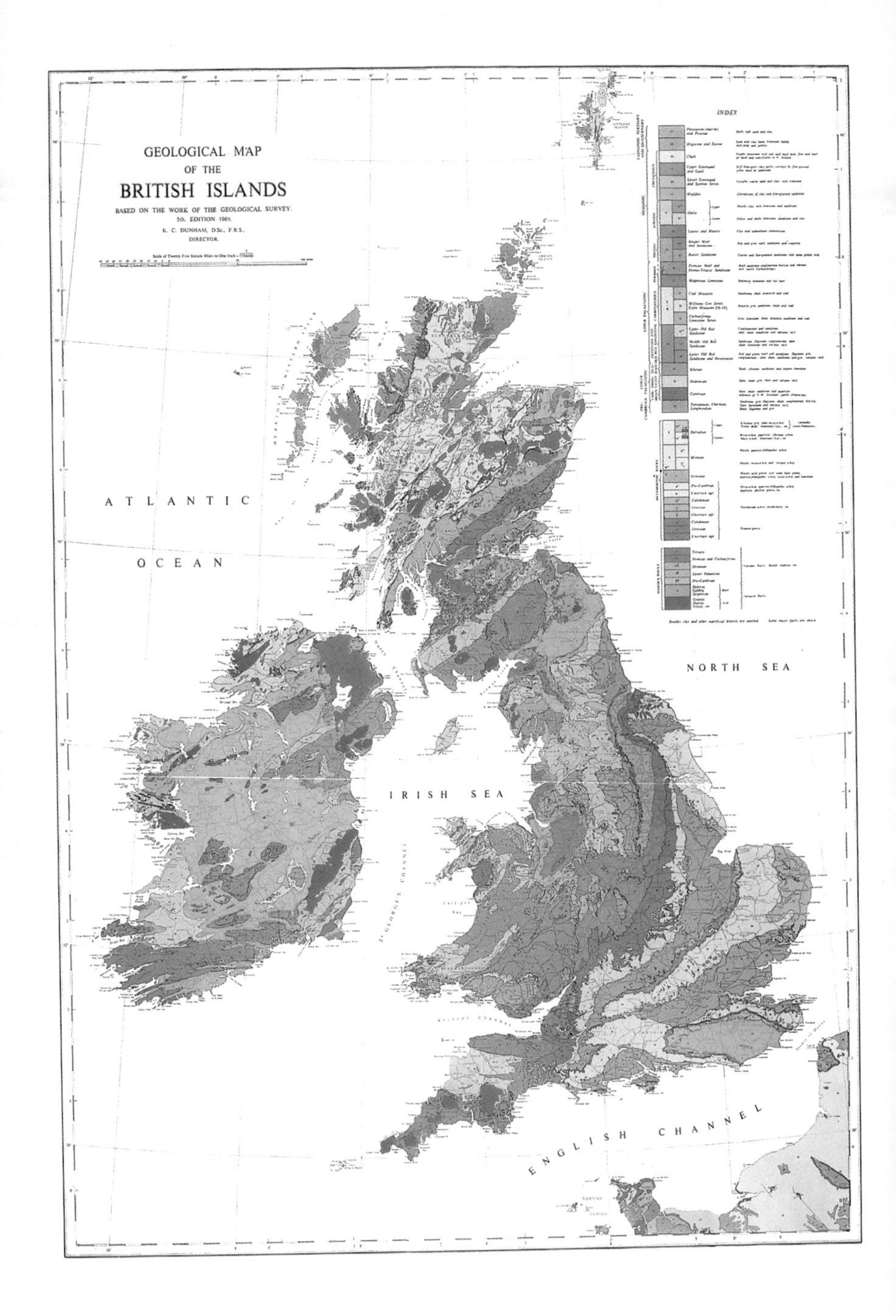
GEOLOGICAL MAP
OF THE
BRITISH ISLANDS
BASED ON THE WORK OF THE GEOLOGICAL SURVEY.
5th. EDITION 1969.
K. C. DUNHAM, D.Sc., F.R.S.,
DIRECTOR.
INDEX
ATLANTIC
OCEAN
NORTH SEA
IRISH SEA
ST GEORGE'S CHANNEL
ENGLISH CHANNEL

10

The Great Map Conceived

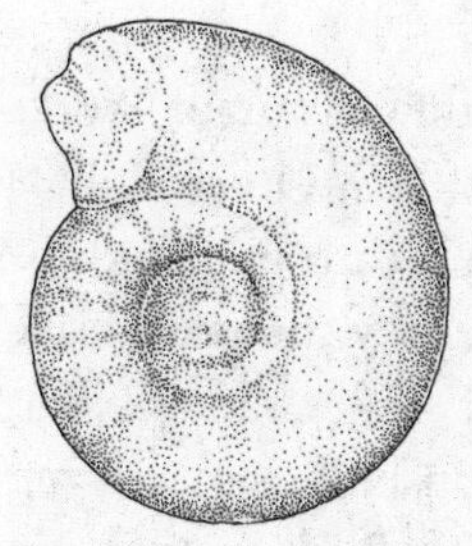

Tulites subcontractus

Each age has a set of defining reference books, that small collection of timeless volumes that is deemed essential to any household that considers itself intelligent and civilized. Today, it might include *Britannica*, the *OED*, or *Brewer's*, with, additionally for the more specialized mind, perhaps *Wisden*, the *Statesman's Yearbook*, or the *Cook's Continental*. A century ago the moderately curious household might have had on hand volumes of *Whitaker's*, *Pears*, or *Cruden's Concordance*, and most certainly a dog-eared edition of a *Bradshaw*. In the eighteenth century the choice would necessarily have been more limited. One would certainly possess (after its midcentury publication date) the two-volume edition of Johnson's *Dictionary*. And if one ever contemplated taking a post-chaise and venturing from home, it would be essential also to have access to a copy of the best-known atlas of the day, what was widely known as a *Cary*.

John Cary, little known today beyond the rarefied world of the antiquarian map dealer, occupies an ineradicable reputation as the leading mapmaker of the eighteenth and early nineteenth centuries. He was only twenty-four when he created his first engraved plan, in 1779, a starkly beautiful, rather plain design that offered something very different from the gaudy, curlicued, and highly confusing maps and charts that were the fashion of the day. His work proved popular, so much so that by 1787, when he produced his first map collection, *The New and Correct English Atlas*, it became the essential reference volume that households were to possess well into the next century. Servants in London were commonly sent out to Hatchards in Piccadilly, their masters demanding simply that they *collect a Cary.* So many editions were produced of the *Atlas* that the printing plates had to be reengraved more than once, so worn with demand did they become.

The creator of these elegant and highly accurate works achieved a towering reputation. The General Post Office (GPO) was one of the first to recognize John Cary's cartographic brilliance: In 1794 it commissioned him to make a comprehensive survey of all main roads in England and Wales, a task that required him to walk the length of each, pushing a measuring device—a stick with a four-inch wheel at the business end—in front of him. The resulting book—*Cary's New Itinerary*—was published in 1798.

It was while Cary was working on the GPO commission that he and William Smith first met. It would be pleasing to imagine that they encountered each other out in the field—the Oxfordshire blacksmith's son, hammer and acid bottle and compass in hand, meeting the Wiltshire maltster's son, with his measuring rod and notebook. Perhaps they would have met somewhere near Midford, where the Fosse Way (which was being surveyed by Cary) crosses the coal canal (which was being surveyed by Smith). But it seems more likely that they actually met in London; perhaps they did so during the 1794 parliamentary inquiry that resulted in the grant to the canal company of the right to begin excavation.

But whenever and wherever it was that the two men first met, Smith and Cary began in 1794 in Somerset a collaboration that was to last, happily if not necessarily profitably, for much of the next forty years. It was a collaboration that began with the quarter-inch survey charts of the canal itself, which were engraved by Cary (the canal company's obvious choice, since he was so well known and well regarded) and had engineering and some geological details superimposed by Smith;* it culminated in publication of the great national geological maps and atlases that lie at the heart of this story.

Wisely, or perhaps timorously, William Smith decided to begin his efforts to map the entire country in a relatively small way. By this time in his life he had a fair knowledge of the countryside and the geology of the counties in which he had worked, either for the canal company or for Edward Webb: He can fairly be said to have known, well enough to make a stab at some rudimentary geological cartography, the lie of the land in Oxfordshire, Warwickshire, Worcestershire, Gloucestershire, Wiltshire, and Hampshire.

He knew very well indeed the deep geology of northeastern Somerset. He had been to the capital many times, and he could tell at a glance, by the dramatic gradations (dramatic to him alone, perhaps) in scenery, soil, and vegetation, the progress he was making from the Lower Jurassic to the Upper Cretaceous

*One of these surviving maps, rediscovered in 1973, shows the precise location of one of the more mysterious engineering marvels of the canal, the huge canalboat lift, or caisson, that was built instead of a flight of locks outside the village of Combe Hay. It was basically an enormous water-filled iron box, mounted on four giant geared legs: The coal barge sailed into the box, which was sealed at both ends, cranked ponderously up sixty feet to the level of the upper reaches of the canal, and the barge released into the higher level. It took an extraordinary feat of technological *brio* to make it; and it was an extraordinary tragedy that it never really worked, was abandoned, and was eventually replaced by a time-consuming, water-using set of locks (until, that is, the canal was replaced—and ruined—by a railway line). Industrial archaeologists for years searched for evidence of the caisson, buried under two centuries worth of mud and neglect: the Smith-Cary map found in 1973 shows its precise location. It was known that a chestnut tree had been planted on the abandoned site some time in the early nineteenth century: it still exists today, and at the very spot indicated by Cary's venerable map.

each time he journeyed by coach from the Circus in Bath to Charing Cross in London. He had been to Newcastle-upon-Tyne, too, and had a passing acquaintance with the underpinnings of Rutland and Lincolnshire, Northamptonshire, Yorkshire, Durham, and southern Northumberland. Enough knowledge, in short, to have a tentative go at making a map of the whole country, albeit on a very small scale.

He chose as his base map a copy of, not a real map at all, but John Cary's index sheet to the second edition of the seventy-sheet *New and Correct Atlas* of England and Wales, published in 1794. On this sheet, which displayed Britain up to the Scottish borders at a scale of about forty-seven miles to the inch (and on which it is possible to see the superimposed numbers of the sheets to which the index refers), Smith colored in what he imagined to be the extents of a number of geological formations—the Tertiaries, the Chalk, the Coral Rag and Carstone, the Oolites, the Lias, the Red Ground, the Magnesian Limestone, the Coal Measures, and the Carboniferous Limestone.

On his map and in his notes he gave names to those rocks that did not already have the dignity. The Cornbrash, the Forest Marble, the Lias still survive—"a system of names almost barbarous to ears polite," wrote one critic. Years later, when the honors began to cascade down on Smith, a distinguished professor offering him a medal was tactless enough to remark on the "uncouth" names he had given to the rocks. Yet some of the more barbaric and less couth names have gone, mercifully: The horizon that is now known as the Oxford clay was in Smith's notes called the Clunch Clay, surely one of the least agreeable rock names that can be imagined.

He worked on his map through the early months of 1801, coloring it in with the information he had gathered on his stagecoach and post-chaise and walking journey, using the same system of coloration he had initiated with the circular Bath map—though as he was now dealing with six or seven formations, not just the three he had colored before, so he had to be more chromatically creative. He chose gray bands for the Tertiary outcrop, blue-green for the Chalk, brown for the Coral Rag and Carstone, yellow for the Oolites, prussian blue for the Lias, and red, not surprisingly, for the Red Ground.

In addition to the simple fact of coloring the strata (admittedly most boldly so), he also incorporated a further device that, he felt, would make an even greater impression on the eye. He shaded each of the colors away from the lowest part of their outcrop—so that at the base (as in the Oolites, where they rest on the next lowest, or next most westerly bed, the Lias) the color is strong, and fades slowly away until the junction with the rock stratum above it, which is similarly strongly colored and fades away. So the Lias is dark blue at its lowest point, faded to pale blue; there is an inked-in line and suddenly there is the Oolite, bold yellow and joining starkly and suddenly with the outcrop below. It then fades slowly away until, after a further inked-in line, there, in a deep chocolate brown, stands the Coral Rag, which fades in its own turn until it encounters the vivid blue-green of the Cretaceous, and so on. The technique was time consuming and is not used today: That Smith decided on it suggests that, with a limited budget for colors and plenty of time on his hands (in the aftermath of losing his job with the canal), he did his best to make his map dramatic looking, even if it was not the kind of technique that would ever lend itself to mass production or publication.

He completed his work by writing, in his best copperplate, the map's title—adding to Cary's engraved phrase "General Map" the words "of Strata in England and Wales" and signing the work "by W. Smith, 1801." It all looks more than a little rough-hewn, embryonic. As indeed it was: This map was only a sketch, a cartoon for a major cartographic task that Smith knew he was not yet up to attempting.

He made at least two further experiments with small-scale

maps later in the same year. Each one was a little more advanced, with Smith either remembering new details, or finding in his notebooks fresh data, or hearing from Richardson or Townsend or his other amateur geologist friends some new piece of information, or being given some new fossil sample or some new piece of outcrop, with a good fresh edge, weathering-free, and from a known and fixed location—all information from which he might make some useful identification* that would further enrich the information on the chart.

Perhaps the best of these three 1801 *ur*-maps—which can surely count as being Britain's first-ever national geological maps, and thus the first useful such documents made of any country, anywhere—is the one which Smith formally presented to the Geological Society of London in 1831. It hangs there still, honored but essentially forgotten—for, though very old indeed, it is much duller in aspect and very much smaller and less distinctive than the giant map of 1815, which hangs nearby and for which Smith is more deservedly famous.

The small map at the society's headquarters does not use a Cary sheet as its base, but a rather larger-scale chart (about thirty-seven miles to the inch) of England and Wales that was taken from a world atlas published by one of Cary's rivals, Robert Wilkinson. On this Smith has colored seven strata: the Chalk is in green, the Coral Rag in purple, a stratum that he calls the Clunch Clay is colored here in a gray wash, the Oolite Freestone in its now-traditional yellow, the Lias–Carboniferous limestone in blue, the Red Ground in red, and finally a hodgepodge of pre-

---

*A common bane of most geologists' otherwise pleasant existence is the person who, all too commonly, offers up a beach stone with the request that he or she *please identify it*. It is all but impossible to hazard a reasonable guess at what a smooth, well-weathered, and near-spherical piece of rock might be. If it is freshly broken, and the new-fractured edge inspected with magnifying glass and tested with acid, a tentative ID might be made. Grinding out a thin section of the rock and using a microscope with a polarizing filter to view its minerals would make its naming even simpler and more certain. But most people on beaches do not want their finds to be broken in half or sliced to a tenth of a millimeter: Better, then, that such gatherers take their trophies home unidentified, to remain—like most rocks and the processes that made them—something of an enigma.

Carboniferous strata that would now be equated with the Old Red Sandstone of the Devonian (which Smith calls by its Welsh name, *Red Rab*), colored in a burgundy-brown wash.

These early Smith maps may in some ways look—especially to the pedantic-minded critics—rather vague, and in truth on close examination they are not at all accurate. There are mistakes: One limestone is confused with another, areas (like the Weald, say, or the North Yorkshire Moors) are left tantalizingly blank, or colored like rocks that don't outcrop within fifty miles.

On the other hand, step back for an instant, and the three maps suddenly look utterly remarkable. They do so principally because they display a pattern, simplistically reasoned maybe, crudely executed surely, but a pattern that is boldly representative of the direction and outcrop of England's main clutch of middle-age sedimentary rocks, and portrayed in a way that has been confirmed time and again in the two succeeding centuries since their publication. Set a copy of a Smith map of 1801 alongside a British Geological Survey map of 1979, and the pattern looks just the same: The underground of the nation is shown in a broad outline that has hardly changed at all, much like the unvarying outlines of the overground.

Viewed from this perspective William Smith's first national map is quite uncannily accurate, and an astonishing achievement. Long swathes of color sweep northeastward up from Dorset to Yorkshire, from Portland Bill to Flamborough Head, from Bath to the Humber, displaying almost precisely where the chalk, the oolite and the Triassic marls outcrop across the heartland of England. The fact that one half-educated Oxfordshire yeoman, working alone—with compass and notebook and clinometer and an abiding appreciation of the beauty and importance of fossils—could surmise with such accuracy what a thousand surveyors and professional geologists have in the decades since really only succeeded in confirming, is little short of a miracle.

✣

Yet at the same time as this triumph, the two scourges that were to afflict William Smith's life—penury and plagiarism—were beginning to signal their distant presence. Since his dismissal by the canal company he had already endured nearly two years without formal full-time employment—although he was by now charging two guineas a day (and sometimes three) for his freelance services, and so was still making a superficially decent living. But frustration—a slowly developing curse—was beginning to become the leitmotiv of his existence: And the first hints were emerging that he might be rather less appreciated than his achievements seemed to warrant, that while he might well be a prophet of a brand-new science, he might also be denied the full honor that prophecy generally deserves.

He was first warned in May 1801 that others might be onto his ideas, and might steal them away. His old friend Benjamin Richardson was the first to make his concerns known, the first to stir Smith into a brief period of literary enthusiasm. Richardson himself was first set to wondering by his colleague Joseph Townsend, who had idly remarked that he thought he himself might write a short treatise on his own huge collection of fossils. If Townsend could do such a thing, Richardson suddenly thought, then perhaps some less scrupulous men—maybe numbered among those who had been so generously sent details of Smith's thinking—might also be drawn to publishing, and to stealing Smith's glory for themselves.

Richardson heard that Smith was passing through Bath on one of his countless freelance excursions, found out that he was putting up at "the Pack Horse, in the Market Place," and penned him a hurried warning, a document that later turned out to be as prescient as its language was orotund. "My dear friend," he wrote:

> To prevent the first admission of the ideas of your communication being turned to another's advantage (which however I

cannot injure our friend the Rev. Jos. Townsend by supposing *him* to have entertained), I assured him before he left Bath that you had determined instantly upon giving it to the public yourself, and that you meant to publish it. . . .

It may be worth pursuing for several reasons: 1st. The printed proposals would secure the discovery for yourself. 2ly. It might be an eligible means of gaining time to go on progressively as your knowledge increased. 3rdly. It would make some returns for the expense of publications as you proceed. 4ly. It would make the work most perfect.

There was more in this vein—but the intention was clear. Someone—and yet most decidedly not the worthy Joseph Townsend—was onto William Smith's ideas and might well write about them and claim them as his own. It stimulated Smith to action within the month, as we shall see; the dangers it suggested were becoming more evident by the day.

It is possible to speculate at this distance that the plagiarist-in-waiting was that other Bath clergyman, fossil collector, and by all accounts, colossal bore* Richard Warner. The man had already been in hot water with the architect John Carter, who accused him of stealing a print of one of his engravings and using it without permission or acknowledgment in a book of his own: He was fined twenty pounds and ordered to pay more than three times the fine in costs. He got into trouble with critics: Once, after writing a two-volume work of *Literary Recollections*, he was dismayed to find that a reader had himself published a twenty-one-page monograph listing all the book's mistakes. And the printers did not smile on him either: His was the book *Topographical Remarks Relating to the South Western parts of*

*He gave long and windy sermons until his retirement, whereupon he embarked on a writing career with books like the encouragingly titled *Nugae Politicae: Solitary Musings on Serious Subjects, by an Aged Man* and the almost equally unalluring *Diary of a Retired Country Parson, in Verse.*

Hampshire that was delayed by a fire at the engravers', which melted all the plates into an immense ball of copper alloy.

But the Reverend Warner's *History of Bath* did appear in 1801, with no outward trouble hampering publication. Not a few eyebrows were swiftly raised when it was noticed that the book included a copy of William Smith's "Table of Strata"—the document that the geologist had dictated after dinner at Townsend's house two years before—incorporated into Warner's book without any indication of either permission sought or payment made. Most probably Smith would not have minded—Townsend himself noted "the open liberality peculiar to Mr. Smith," and recalled his desire "to make so valuable a discovery universally known." And in any case, it was making him more widely known, which for a freelancer can have been no bad thing.

But ten years later still, Warner came out with another book, and with it another purloined map—this time a direct copy of the circular map of Bath that had been made by Smith in 1799. Warner had this time renamed it, inelegantly, "A Fossilogical Map of Bath and Its Environs." It was crude but useful, in that it showed all the local villages and beside each name, the stratum that was most commonly evident there. It was a naked example of plagiarism. Smith never complained: No doubt he found it too vulgar a creation to raise objection.

He would make a very great fuss, though, in later years, when still others stole his work. "Men of scientific eminence," he would later write, famously and scathingly, in a letter to a friend, were all "pilferers of information," who saw it as their right to regard all unpublished observations as "lawful plunder."

In due course—fifteen or so years following this first brush with a small-time plagiarist—such pilfering and plundering, though on a far grander scale than the Reverend Warner's, would help consign William Smith to debtors' prison and to years of homelessness. It would leave him embittered toward

Londoners, toward the city's intellectual and social elites, and toward those in the science who, he felt, rarely ventured out of their drawing rooms, rarely dirtied their soft pink hands, rarely muddied their fine leather boots. In 1801 he did not see in Warner's peccadillo any indication of the sorrows to come. He was at thirty-two still something of an innocent, successful and of sunny disposition, and the world seemed a kindly place. But it would not be too long before all this would change.

He was by now fast building up an impressive circle of friends and was winning commissions that would take him clear across the British Isles, satisfying his goading urge to travel, to keep moving. The Bath Agricultural Society was the key: Its membership reflected the extraordinary reputation of the region for comfort, beauty, wealth, and style, and it exposed Smith to a range of men of money, leisure, divinity, and science, as well as to men of the same kind of rudely practical bent as himself.

Each group of men was peculiarly useful to him. The gentleman amateurs, the fossil collectors, and the natural philosophers who invited him to their vicarages and country cottages encouraged him in his studies and helped him with his ideas. The practical men, the coal borers and drainage engineers and well sinkers he met, taught him new techniques and allowed him to make the best use of his time in the field. But it was really the aristocrats, the members of the landed gentry, who at this stage in his life were to provide him with both the work and an entrée into ever-widening network of influence.

There now came a sudden acceleration in William Smith's ambitions and desires, and central to this change were the brief friendships he enjoyed with three famous men of the day: Thomas Coke of Holkham, the duke of Bedford, and Sir Joseph Banks. Considering Smith's later belief, when he was imprisoned and in trouble, that it was the English aristocracy that had treated him badly, and that his humble beginnings had counted heavily against him in a society so class-obsessed as nineteenth-century

England, it has to be remembered that the nobility assisted mightily in bringing his early ideas to fruition. His early sponsors were indeed members of the English upper class, writ large.

The chain of introductions to these sponsors was brought about by way of Smith's unusual expertise not with rocks or fossils, but with water.

Canals, with which Smith now had a good deal of experience, are essentially enormous and elongated repositories of water. They are not rivers: The water in them does not have a natural source, does not flow from one end to the other, and is not continually refreshed from a spring. All is artifice, a complex and cumbersome arrangement of planning and engineering that allows a very large body of water to exist in a series of deep horizontal chambers along which vessels may glide, passing uphill and downhill by way of long cascades of locks, in order to move goods, or people, or to undertake commerce.

It is central to the design of any canal that it retain its water as best it can, since water is costly and has to be brought into the canal from rivers or lakes or purpose-built reservoirs. Smith came to know very well—almost uncannily, his admirers said—just how to route a canal so that it lost as little water as possible. He saw to it that wherever practicable it passed over beds of impermeable rock—and in those places where it did not, that its bed and banks were lined and proofed so that the water standing inside stayed where it was.

In accumulating what would later become his nationally known expertise in keeping water where it needed to be, and removing it from where it shouldn't be—William Smith came into sudden demand by farmers, who saw in his skill a way for them to turn their profitless marshes into workable farmland. Up to this point he had been known for his skills as a surveyor and a cartographer; now he was changing, chameleonlike, with the addition of this new and very marketable skill, into the unglamorous but, to postenclosure England, essential figure of a drainage engineer.

Recognition of his growing mastery was one reason why the chairman of the canal company, James Stephens, had in 1799 hired Smith to help him drain his own farmland. No matter that Stephens had fired Smith from the canal that very June: So bad was the rainfall that autumn, and so uselessly boggy did the Stephens family farm become, and so in need of employment was Smith himself, that a deal was struck—in which Smith was paid to drain, dredge, and dike the Stephens fields for the highly respectable rate of two or sometimes even three guineas a day. And Smith, clearly burying his pride, worked well: The Stephens farm was promptly drained, any number of cuts and culverts were made and pipes and bores laid, and the farmland was made ideally suited for agriculture for years to come.

This was the time of the "improving farmer"—of the agriculturist who, now that the enclosure acts had brought some sanity to the fields of England, was intent on making as much as possible from the land he worked, by using newfangled fertilizers, by mixing soils, by judicious draining projects, by breeding new strains, and by sculpting the land and creating new environments. Thomas Crook, a typical improving farmer of the day, who lived in the Wiltshire village of Tytherington, saw what Smith had done to the Stephens farm nearby, and hired Smith to do much the same for him a year later.* And then, once the drainage work was successfully completed, Crook invited for an inspection tour the man who is quite probably still regarded as the greatest agriculturist of his or of any age—Thomas William

*While he was working at Crook's farm, Smith made a discovery of another horizon, that of a friable sandstone containing a fossil called *Sigaloceras*, to add to his slowly expanding collection of strata. He called this one Kelloway's [*sic*] stone, after the numerous road-stone quarries near the village of Kellaways [*sic*]: He placed it in the upper subdivision of what in his 1799 table he had called simply number 4, "Sand and Stone." The horizon was later to be called Kellaway's [*sic*] Beds, and the name of a geological stage, Callovian, was derived from it—the one name that Smith has bequeathed to a major period of earth history. He also gave the name Cornbrash to the limestone immediately below, using the quarrymen's term *brash*, and adding to it the fact that corn seemed to grow well in the fields above its outcrop.

Coke of Holkham, or as the Prince Regent later liked to have him known (since Coke took the title with great reluctance) the Earl of Leicester. For Smith the meeting presented an opportunity of inestimable value.

Coke of Holkham, as he was generally known, was a man with initially no practical knowledge of farming—he simply owned farms, placed tenants in them, and lived off the rental income. But in 1788 one of his tenants refused to renew his lease, and Coke decided that, rather than let the land lie fallow, he would make an attempt to farm it himself. Since he knew so little he took the radical step of organizing a huge seminar, which he called a sheepshearing, and to which he invited all the local farmers and practical men and landowners so that they could inspect his land, crops, and livestock, and make recommendations. The event was enormously useful to him and, since he laid on huge lunches and dinners and had experts offer speeches and demonstrations, great fun for those who came.

In due course his own farm flourished. He experimented with new soils and fertilizers, he replaced the normal Norfolk crop of rye with wheat; he decided to buy and crossbreed sheep, and to introduce into his fields large numbers of sturdy, fat, wool-covered animals that would replace the scrawny, doglike specimens with which Norfolk was then usually populated. He bred Suffolk pigs with Neapolitans, and within two years was producing massive porkers that delighted markets and trenchermen alike. The sheepshearings—known across the land (and through much of farming Europe) as Coke's Clippings—became hugely popular: One of them, held in the early summer of 1818, attracted seven thousand people, with Coke offering hospitality to more than six hundred in Holkham Hall, no matter what their rank, station, or nationality.

Thomas Coke's reputation rests today largely on the outward appearance of his great farms, and on the now-widespread knowledge of his techniques of breeding and feeding. What is not so often recalled, about Coke, nor indeed about any of the other great improving farmers of the day, is that almost all these men, before they sowed a single seed or bred a single animal, had first to prepare their lands.

Before the enclosure acts, English land was in a hopeless mess. Unfarmed, the newly enclosed fields were still merely boggy patchworks of mud and sedge, with barely any meadows suitable for workers to work them with plows and seed drills. Very few of the new estates were unencumbered by piles of rocks or clumps of trees. Fewer still, more important, were properly drained.

The enclosure acts changed all that, by prompting the newly empowered owners to recognize the need for efficiency and careful husbandry, to come to grips with their individual agricultural shortcomings, and to begin shaping the tidy and mannered English countryside that we see today—fields laid out neatly, sedge trimmed back, bogs all drained. The fact that all is so impeccably and memorably attractive today stems in great part from the work of men like William Smith, who were called in by landowners like Coke to change out of all recognition the appearance of their vast acreages.

William Smith was called in to Holkham specifically because the work there was difficult, and because he had evidently been so successful in performing drainage work on the canal, as well as for Stephens and for Crook. In Norfolk he was asked to dry out the huge flat fields that stood beside the North Sea shore, to make productive the salt marshes that lay behind the dunes, to channel a network of wayward rivers, to dredge and drain and otherwise hydraulically improve the lands. His work remains intact today—hundreds of adequately dry and highly productive flatland acres that surround what is now recognized as one of England's most enchanting stately homes.

A year after Smith had accepted the commission, and at about the time he was coloring his geological discoveries on the Cary maps of England, Thomas Coke introduced him to one of those figures of whom Lloyd George was later to be so rudely disapproving—the vastly wealthy and hugely influential Francis Russell, the fifth duke of Bedford. "Deficient in wit and imagination" though he may have been regarded (by Emma Louise Radford, who wrote his entry in the *Dictionary of National Biography*), the duke was a great agriculturist,* and his enormous estates at Woburn—now a three-thousand-acre deer park for tourists, with nine species of especially adorable animals—were made into a model farm along Coke's lines, and supported vast herds of cattle and flocks of sheep.

The Duke's four-day sheep-shearings were so popular as to make even those at Holkham look like village fetes: thousands came, and there were ploughing contests and cattle sales, wool auctions and dances, and banquets for many more hundreds than even the Earl of Leicester could afford. "To see a Prince of the Blood Royal and many great Lords sit down to the same table," wrote Arthur Young,† one of those who went, "and partake of the conversation of the farmer and the breeder; to see all animated in the spirit of improvement, and listening with delight to the favoured topic of the plough, is a spectacle worthy of Britain, and in her blest isle alone to be beheld."

A suitably massive oil painting survives, by the noted animal artist George Garrard, of the great shearing held at Woburn in 1804. Partly a record, partly an allegory, it shows eighty-eight "agricultural personalities" grouped around the base of a massive limestone column capped by the "Ship of Commerce." The duke is there, top-hatted and, suitable to both his *gravitas* and *dignitas*, the only figure on a horse. Around him are the great and the good of the English rural universe: smocked farmers and shepherds, impeccably dressed gentleman farmers, roughly

---

*As well as being a great city planner: He razed the old Bedford House in central London and in its place created Russell Square and Tavistock Square, the former being one of the capital's largest.

†In *Annals of Agriculture.*

dressed blacksmiths and farriers, men fat and thin, jolly and severe, ill-born and noble, of practical or professional appearance, all busy in conversation, or gazing in rapt attention at the tups being shorn in a pen before them, or at the elongate cattle standing patiently at center stage.

Garrard's painting seems to show that, despite this being June, the shearing was held on a wild and cloudy day. The buildings in the background are severe and practical, all stables, byres, and dairies, and none of the Inigo Jones–designed masterpieces in which his grace lived (as a lifelong bachelor) visible. To judge by the trees alone it is a very English painting: One of them is an age-gnarled oak—perhaps the very tree, which still stands at Woburn to this day, where a Cistercian abbot, one Hobbs, was hanged in the sixteenth century for making (according to the court records) "treasonable utterances" against the king. It was his supposed crime that led to his abbey being confiscated and handed to the determinedly Protestant Russell family, which has lived there ever since. Of such events, William Smith may have thought, has England's aristocracy been made.

For Smith was there, and he is included both in Garrard's painting and in an aquatint engraving the artist made seven years later. He is only barely discernible, however—still regarded as only a peripheral member of the ducal elite. He is just visible on the picture's upper left, amid a crowd of others of equal honor and distinction: a bluff-looking man of middle height, wearing a black broad-brimmed hat, looking away, barely recognizable. The artist helps by providing us with a key: his figure is shown as number 10, against which is written the simple and almost vaguely insulting rubric, damnation with the faintest of praise: "Mr. Smith," says the note. "The drainer."

Matters of rank and propriety probably meant little enough to Smith at this stage in his career: He was as close to nobility and power as an Oxfordshire countryman could expect to be. And besides, the links he had forged in the brief time since his removal from the canal company—from James Stephens to Thomas Crook, from the Thomas Coke to the duke of Bedford—were proving both profitable and, as it happened, enormously useful. Moreover, further links in the chain were to be forged through his passing acquaintance with the duke: Not only did he meet and present his card—"Wm. Smith, Surveyor and Drainer"—to still more noblemen, like the duke of Manchester, the earl of Thanet, Lords Talbot and Somerville—but he also engaged his first apprentice, the first man who became a geologist as a direct result of working for Smith.

This was the duke of Bedford's land steward, John Farey—a figure who would champion his mentor's work, play a vitally important role in his affairs, and lead him still further into a world of influence and connection that would enable him, finally, to produce and present the great map that would render him famous.

The story of John Farey (and his own introduction of Smith to one further world-renowned figure who would become the map's greatest and most influential patron) belongs properly a little later in the tale—except in one respect. For when the two men first met, at Woburn in October 1801, William Smith had something to show him.

✠

Benjamin Richardson's stern warning—that if Smith didn't begin committing his thoughts to paper, someone else would beat him to it—had apparently sunk in.

He had indeed embarked on the publication of his ideas. He had accepted the good doctor's advice not to publish merely a new map, or a list of strata like the one he had dictated at Pulteney Street, or even a cross-sectional portrait of the English underground—but a proper book.

And so on June 1, barely two weeks after receiving Richardson's letter, Smith surprised and delighted everyone by publishing a document—a four-page prospectus for the book he was now determined to write. It had a title that, back at the beginning of the century, might have sounded more tempting than it does today: *Accurate Delineations and Descriptions of the Natural Order of the Various Strata That are Found in Different Parts of England and Wales; with Practical Observations Thereon.*

There was a closely formulated financial model: Two thousand copies of the final book would be printed, and they would be sold at two guineas each, with Smith taking a 50 percent cut of the expected profits of £3,200. The prospectus—which Smith showed to an impressed John Farey in 1801—was a suitably handsome creation: Printed in Covent Garden, it was ambitious enough to include a stratigraphically apposite epigraph from Alexander Pope: "All Nature is but Art Unknown to Thee./All Chance, Direction which thou canst not see." The hope was that those who saw the elegant little document would be seized with a burning desire to own the eventual book.

It seemed at first to have the required effect. Letters rained in, all enthusiastically asking to be put on the list. "I have distributed your Prospectus amongst my friends," wrote one Richard Gregory from Coole, in Ireland, quite typically, "and have the pleasure to request you will add to the list of your subscribers the name of my father. Robert Gregory, of 56 Berners-Street, London, and the Hon. Richard Trench, MP, Spring Garden Terrace, London."

Smith had a publisher all lined up, supposedly one of the best. He was John Debrett, who had already made his name and might well have made a fortune as the most noted cataloguer of the peers of the realm. He was Piccadilly's most celebrated Whiggish biographer, a man curiously obsessed with publishing books about Australia and by all accounts an amusing and unreconstructed snob.

So everything thus far seemed set. Even the normally cautious Dr. Richardson appeared delighted, and he wrote an overjoyed letter to his young friend—a letter that reminds us today both of the antiquity of the moment and of the lingering hostility and suspicion felt toward France, so soon after the fall of Napoleon. Richardson advises Smith to:

> take Debrett's opinion on the propriety of giving an edition of the work in Latin for the benefit of all Europe, to be circulated under the patronage of our foreign envoys, etc. etc. This would give the system its due importance, and prevent any pirated French edition, which the world would be ready enough to catch at.

But Debrett, who might well have made a fortune, was not to do so with this particular volume. His finances, spiraling out of control, turned out to be the cardinal problem both for the firm itself and, rather more ominously, for William Smith. For though the celebrated *Debrett's Peerage* was soon to become a reference book of biblical standing among the aristocracy (and always regarded as a considerably more venerable text than its rather *arriviste* rival, Burke's), John Debrett himself was not the man to make money from it, or indeed from any of his publishing ventures. He was described simply as "a kindly, good-natured man but without business aptitude," and he lived well only because of moneys inherited by his wife. He went bankrupt twice; the entire project, along with many others in which he was involved, was soon mired in crisis.

Smith himself set to work with a vengeance. "I have just come off a long and troublesome journey through Somerset, South and North Wales, Chester and Lincolnshire . . . a distance of about 800 miles," he wrote Debrett from Woburn, six months after issuing the prospectus.

> I have scarce had time to sit still, write or rest for the last five weeks, but have picked up much new matter and confirmed

> some of the old, and collected the following list of subscribers along the way—R. Cornfield, Dr. Beales, Mr. John Grant (3 copies), Mr. Wm. James of Bristol . . . [he names a further fifteen potential purchasers, including a surveyor named Jeremiah Cruse with whom he would soon go into business]. It was my intention to come to London almost immediately . . . but the Duke of Bedford has been making such appointments for me in Ireland that I must attend to.

But soon John Debrett was making it clear that Smith did not have a prayer of making anything like the sixteen hundred pounds he had naively anticipated. Smith was disillusioned—and devastated. He sent Debrett an understandably churlish note.

> A work of public utility which has engaged 12 or 14 years of the best part of a man's life in the most laborious train of thought and observation ought to close up the past prospect with plenty. Remote expectations will not satisfy me any longer. I am certainly under no obligation to lay before the public the fruits of my labour and unless I can be satisfied of an adequate reward I cannot think of giving up more of my time.

Although money was the most visible problem in the making of the book, it was not the only one. Smith's lack of literary confidence contributed too, as did his curious indecisiveness, his seemingly pathological need to procrastinate. It was to become a lifelong affliction.

On some days he would decide to abandon the whole affair: "Making little progress from not being able to please myself in the mode of expressing my thoughts," he confided to his diary, "and from foreseeing great difficulty in arranging such matters for a book, and also from considering that I was losing two guineas a day for the chance of a small profit by the book, I wisely decided to stick to my profession." Yet on other days he would seem reenergized, eager to start afresh: In May 1802, for example, we find pages from his daybook recording a number of small victories over his indecision:

> Tuesday May 11: wrote several pages on the formation of the strata and the effects of the deluge &c. . . .
>
> Wednesday May 12: collecting together loose memorandums . . . writing several small pages. . . .
>
> Saturday May 15: considering about plates to be engraved . . . looking over and sorting out maps. . . .
>
> Sunday May 16: After breakfast wrote eighteen pages of observations. . . .

Yet it was not to be. The relationship between Smith and Debrett deteriorated fast, with the geologist asking far too much, the publisher offering far too little, the project becoming subject to interminable delays and snared in arguments between the pair. In the end it was the money that did it in: Debrett went bankrupt in 1802 and then again in 1804, and Smith's hopes of having his work offered between hard covers to his enormous list of subscribers, or to the public at large, swiftly died. The book was never to appear. And Smith was not to write anything of geological importance for nearly a decade.

But the setback then seems to have infused him with a new but very different sense of purpose. Although he had been forced to abandon his writing, his new circumstances served to provoke him into a new frenzy of travel. All of a sudden he was accepting commissions throughout the length and breadth of the country—Norfolk one week, Dorset the next, Yorkshire today, Shropshire tomorrow, and, with the duke of Bedford's ready help, Ireland too. He began a period of intense restlessness, burning up the stagecoach miles like a traveling salesman, seeking out the work, and at the same time seeking out the rocks and fossils that unrolled and unraveled themselves before him.

✠

The notion of publishing *something* still nagged persistently at him like an aching tooth. Maybe it should be a book, or maybe it should be something far grander, far more ambitious—maybe some document that demanded less intellectual energy, less cerebration, but that could perhaps emerge as a direct consequence of all his wandering, his collecting, his fieldwork, his observing. Maybe, if it was not to be a book, then it could and should be a truly wonderful, majestic, all-encompassing map.

And thus, by circumstantial happenstance, was the plan for the great map formally conceived. It was an idea that had nudged at him for years, ever since his first youthful attempts in Bath. The maps he had made in those early days were rudimentary enough, either devoted only to discrete localities or performed over a larger scale with a broad-brush vagueness that made them relatively simple to complete. But now it began to seem to him that a grander, more ambitious map could properly memorialize him—a map of the whole country, closely detailed and highly accurate. This new map, he thought, would reflect, in so much more appropriate a manner than a book, his special talents.

To make such a thing would require a great deal of time and a lot of money. And so he set about to gather the funds and rally yet more public support for this new project, and at the same time accelerated once more his determined wanderings. He traveled on commission, draining, surveying, mapping—and gathering as he did so yet more and more information.

"I intend to come through a part of Hampshire on my way westward," he wrote to one Samuel Collett in May 1802, "for the purpose of seeing a few places where the chalks and clay strata turn round the end of your hills." And to Richardson back home in Bath:

> I have collected a great deal from the North of England and Scotland. Our Mendip limestone, with St. Cuthbert's Bead,

goes out to sea at Holy Island, where they are found in great plenty, and are called by this name from the saint of the island. I have found fossils in red marl of Staffordshire, connected some limestones, and nearly connected some ranges of the coals.

It was all information that would soon enable him to sketch in his mind, and with steadily accumulating detail, an outline of what he now surely wanted to produce: a definitive portrait of the immensely complicated arrangements of the strata, a chart of the underworld of his country.

II

A Jurassic Interlude

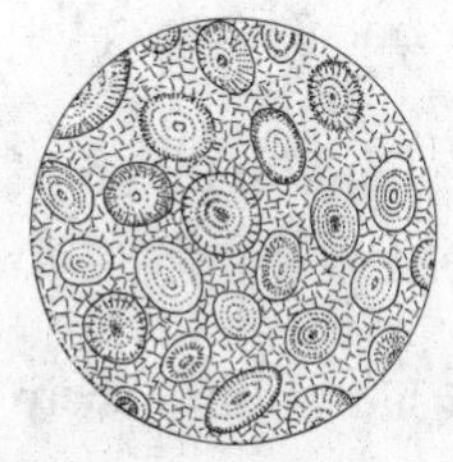

A magnified cross-section of a typical oolitic limestone

On early summer Saturdays in the 1950s when the weather was fine, the sisters of the Blessed Order of the Visitation liked nothing more than to take their little convent boys swimming in the sea. To keep us more biddable they tried to tire us out by walking the entire way, taking us up and over the little range of steep Jurassic hills that lay between school and the beach.

I think that none of us appreciated much of the geology at the time; and nor, I daresay, did the nuns—but we all knew that the landscape it formed was unusual and very beautiful. And although my own memories of those long, long walks have been more than a little colored by the harshness of the boarding-school regime (I was six and a half; and the punishment for my frequent dawdling on Saturday walks was six strokes on the hand with a bamboo cane, and a big wooden spoonful of castor oil forced down my throat by an unusually ugly sister who sported a luxuriant blond moustache), there are delights about it that linger yet.

There was a simple routine to the excursions. We would set out from our convent in the old rope-making town of Bridport, in Dorset, and walk for a while beside the main road that took westbound travelers on toward Devon and Cornwall. A mile or so out and we reached a village called Eype, where we would turn south and stop to rest and play games on a sloping meadow thick with ferns and yellow gorse. Then we would plunge off again uphill, into a rabbit warren of steep lanes that had been carved into small canyons of a honey-colored sandstone, and that were concealed from the landscape beyond by hedges six feet high, with a feel about them like front-line trenches—that of course we as brutish little boys very much liked. Occasionally there would be a five-barred field gate, and through it tantalizing glimpses of stands of rolling wheat, warm and golden in the early summer sun.

The roadway in these lanes would be littered with small pieces of rock that had broken from the canyon walls. We would kick them around, imagining them to be grenades or spare ammo, and we would scuff our shoe leather white on their sharp edges, until a disapproving escort nun wagged her finger and reminded us of the probability of public punishment in the school refectory the next morning. The list of offenses fit for the caning-and-castor-oil routine was very long.

After fifteen minutes of slope that had us puffing and panting upward, the road flattened and widened, the canyon walls fell away, the summit meadows appeared, and ahead was the vision that we had all awaited—the sun-dappled hammered-pewter surface of the English Channel—the sea. Up here there always seemed to be a cool onshore breeze blowing up and over the summit. It was tangy with salt and seaweed, and the way it cooled the perspiration was so blessed a feeling that we would race downhill into it with wing-wide arms, and it would muss our hair and tear at our uniform caps, and we would fly down toward the beach and to the surging Channel waves that chewed back and forth across the pebbles and the sand.

I seem to remember that by this point in the weekly expedition the dozen or so of us—all called by numbers, since the convent's peculiar regime forbade the use of names; I was simply 46—were well beyond caring what the nuns might think: The ocean was by now far too magnetic a temptation. Once in a while we might glance back at them as they stood, black and hooded like carrion crows, fingering their rosaries and muttering prayers or imprecations—but if they disapproved of us tearing off our gray uniforms and plunging headlong into the surf, so what? This was summer, here was the sea, and we were schoolboys—a combination of forces that even these storm troopers of the Blessed Visitation could not overwhelm. So we stripped down to our trunks; we paddled, we swam, we splashed and fooled around in the green water and spray for what seemed hours, shouting ourselves hoarse, forgetting everything in our careless summer delirium.

And then, all too quickly, the heat of the day fell away, and it became late afternoon. The sun was suddenly lower, the shadows longer, and in the sea breeze was the first indication of the evening chill, prompting the occasional shiver and excited comparisons of goose pimples. The sea would be enveloped in a pale purple haze. By now, all weary from swimming, we would retreat some yards from the seaside and begin to play in the still-warm sand. We would build castles, collect starfishes or dogfish egg sacs or shells, or bury one another in tombs that would fill coldly in the slow-rising tide.

It was during one of these lazy after-swimming reveries that I found my first Jurassic fossil. It was an ammonite, coiled in repose like a small and fat spiral spring, the relic of a shell that the textbooks said once held a squidlike animal that could pulse its way slowly and silently backward through the warm and life-rich seas of the time. Although I have long since lost it, I remember it only too well. In my mind now it seems a much-loved talisman of those curiously contented days.

The fossil was a smooth, reddish, circular object that sat nicely in the palm of my hand, weighing perhaps a quarter of a pound. It was smooth enough, I thought at first, to be used as a skimmer on the flat surface of the sea. It was almost perfectly whole, not much bigger than an old English penny piece, maybe an inch and a half across, and in comparison with other ammonites I had seen in museums, it seemed to have almost no coils at all—its circumference was simply a circle that folded into itself and then vanished. In its center was a small whorl; and its flanks were traced with sinuous lines, presumably marks that showed how its shell grew, season by season or day by day.

I gazed closely at it, enraptured by its strange delicacy: I licked it to remove some of the sand, and used a fingernail to try, unsuccessfully, to pry away small concretions—these would yield only to the point of a nail file later in the evening. A few of the other boys seemed interested—I remember still that numbers 6 and 25 in particular had shared my fascination and had asked to have a closer look.

The beast evidently left as much of an impression on them. Many years later I came across number 25—by then he had a name and was a senior partner in a private banking firm—while walking along Connaught Road in Hong Kong. It was during an evening of reminiscence some while later that he asked me if I still had that pretty little ammonite? But no, I said shamefacedly, I didn't; and neither of us could remember much more about the day it was found, nor, to our greater shame, could we remember what 6, the other boy who had liked it, was really called.

All this I may have lost or forgotten—but I still remember exactly where I found the fossil; and, knowing that William Smith had been to the very place himself, sometime during the fifteen years that he wandered around England exploring and making ready the details of his great map, I decided that I had to go back there. I thought it would be helpful to make my own brief tour of the Jurassic, to follow the outcrop of England's most distinctive raft of rocks just as Smith had done, from coast to coast, all the way up to Yorkshire from down here among my school memories, in the magic depths of South Dorset.

Maybe by doing so I would come to learn a little more—learn not simply about the topography of Middle England, nor of the lithologies of the Middle Jurassic nor even of the paleogeography of the era—but about William Smith himself. Perhaps if I took the journey he had, I might gain some clue as to the extent of his achievement. And though I would never quite feel it as he had done—I would never quite know the discomfort of bouncing in a springless chaise on a rock-strewn turnpike, or suffer the cold comforts of a windy coaching inn, or the misery of arguments with an undertipped ostler, still I might come to feel just a little as he had done, all those years ago.

And I had it in mind that, despite the newness of all that now rose on top of the landscape—despite the motorways and power pylons and cell-phone masts, the new cities and New Towns and the forests of skyscrapers and suspension bridges and landmarks that had been built in styles and for reasons that could barely have been imagined two centuries ago—there would be something about the outcrop of the period that would be discernible still. The imprint of geology has an immense power over landscape: the imprint on England of the Middle Jurassic would, I imagined, still be there, underlying everything, imperturbable, immovable—and quite probably, just as in William Smith's time, instantly recognizable.

✢

Down at the starting point, forty years on—and the village of Eype still looked much the same. The main road was a divided highway now, with traffic thundering endlessly to and fro along it, and not a line of schoolchildren—who would have been in great peril from the frantic lorries—anywhere to be seen. The lanes at the top of the Downs were just the same, though—

sleepily unpopulated, and incised into the sandstone so deeply that one still half expected duckboards and mud, and the thud of wartime shellfire. And then from on top of the hill, sighing and slumbering below, was the English Channel, just as it always must have been, a soft surging sound on the gravel shore, and with distant ships crawling along its steel gray surface, silhouetted in the southern summer sun.

I walked down to the beach, and, remembering just where I had found the little fossil, turned promptly left to walk the few hundred yards to where I might discover another. In turning left, I was now facing east. A map would show something of a geography that my years in Dorset had allowed me to know quite instinctively. The counties of Devon and Cornwall ranged behind me. Brittany was invisible a long way off across the sea to my right. The rich farming valleys of Thomas Hardy's Dorset were up above me, where I had come from, to my left. And, most significantly to any seeker after geological truth, unseen but very clearly marked on any map, the White Cliffs of Dover lay a hundred miles away, directly ahead.

And in that single topographic fact is a clear and present indication of the simple existence of geologic time. The white Dover cliffs are made of chalk, which is the best-known rock of a period*

---

*As the illustration on page 302 suggests, there are nine basic divisions of geologic time. The division of greatest chronological extent is that known as the eon—of which there are commonly thought to be four, and of which the most recent is the Phanerozoic. This particular Eon is then divided into four eras (the Mesozoic) some of which are divided into suberas; the eras into periods (the Jurassic) and in a few cases subperiods, and the periods into epochs (the Lias). Below this—divisions that may be only five or ten million years in extent, a blink of a geological eye—are a vast number of stages (the Toarcian); and below this, named after the dominant identifying fossils that are to be found only in that particular portion of time, if the rocks and thus the conditions of deposition are of a certain kind, are the hundreds of geological zones. Below there are divisions of mere tens of thousands of years—and thus there are thousands of them—known as subzones (or, by some, as teilzones or epiboles). The rocks below the very top of the East Cliff at Dorset's West Bay—the one in which I found the smooth and barely coiled ammonite—are of the Mesozoic era, of the Jurassic period, of the Liassic epoch, and of the Toarcian stage. The identification of the ammonite itself, as we shall see, determines the exact geological zone.

that, identified by the fossils it contains, has long been known as the Cretaceous. The rocks here in this part of Dorset, on the other hand, were all—and have been similarly identified by another group of very different fossils—laid down during the period known as the Jurassic.

For anyone today to walk eastward from Dorset to Dover along this coastline, just as William Smith had walked eastward along the Somerset Coal Canal from Dunkerton to Limpley Stoke some two centuries before, is to walk forward in geological time—is to walk away from and out of the older rocks and toward and into the newer. The cliffs that ranged before me now were each made of rocks that were successively younger than those in the cliffs that ranged behind me. The more distantly ahead of me they ranged, the younger and the younger they became—so that those lost in the shimmering haze of the afternoon belonged to whole stages and epochs of geological time that were far more recent than those beside and behind me.

It was the gentle and uniform southeasterly dip of all these outcropping rocks that made this possible: Had the rocks not dipped at all, but remained horizontal all the way along the coast, the outcrop would be entirely the same, the topography unchanged, the crops above and the view below unchanged from Dorset to Kent. It was the dip that allowed the history of the underworld to be on such dramatic display—the same gentle and uniform dip that had made all this history so suddenly clear to William Smith in the hills and valleys around Bath.

As the cliffs rose and fell in ranks along the coast in front of me, I could see with consummate ease exactly how each one, because of the bread-and-butter arrangement of its rocks, was geologically different from another. Here beside me the cliff might be made of a Jurassic sandstone capped by a limestone; but two miles further eastward down the coast the same limestone that here was *on top* and formed the cliff's cap might well form *the base* of the next range of cliffs. Layered on top of it there would be other rocks—shales maybe, or marls, clays, ironstones, perhaps more limestones, more sandstones—and with each stratum on top younger and younger than those below until, a hundred miles away, way past the counties of Hampshire and Sussex and well beyond the towns of Bournemouth and Southampton, beyond Brighton and Eastbourne and the great promontory of Beachy Head, there would be the White Cliffs themselves, the chalk.

The rocks here in Kent could be shown by their fossils to belong not merely to a younger geological stage or a younger epoch than those back in Dorset. They would belong to an entire geological period, the Cretaceous, that was a full fifty million years younger than these Jurassic rocks that stood beside me. To cross the southern coast of England, west to east, is thus to travel forward—and at breathtaking chronological speed—in a self-propelled time machine. With every few hundred yards of eastward progress one passes through hundreds of thousands of years of geologic time: A million years of history go by with every couple of miles of march.

Here in this particular corner of Dorset, at the coastline by the tiny port of West Bay—which is where we as schoolboys frolicked in the sea—one huge cliff made of sandstone and topped by a cap of limestone dominates all the scenery. It stood to the east of the tiny harbor and was known, somewhat unimaginatively, as East Cliff. And it was exactly here, cold and swim weary back in the 1950s, that I found that first ammonite. Some forty years on and I had walked down to the sea once more,* and I had turned to face eastward—and the selfsame cliff rose dramat-

*I had come back largely on the advice of a man I had met in the village of Chideock, where I was staying. He was called Denys Brunsden, and he was known by the locals simply as someone who was interested in the local geology. He turned out to be an internationally renowned expert in the physics of landslips, was summoned by countries all over the world—and, when I told him the subject of my interest, dashed out to his study and returned with his most recent award: It came from the Geological Society of London, had

ically ahead of me. It looked quite unchanged—solemn, brooding, and immense.

I walked over to its base, past the bundled-up vacationers and the few pet walkers braving the late spring chill. In part I hoped I might find another fossil; mainly, though, I just wanted to gaze up at what must have enthralled William Smith as he contemplated the beginning of his fifteen-year mapmaking tour, as he reached the starting point of his thousands of miles of trekking along and across the outcrop of the English Jurassic.

East Cliff rose up in a great domed cathedral curve, rising from where two small streams had eroded it down to sea level on its western and eastern ends. At its center it was maybe 130 feet high, vertiginous, and quite sheer. Its face was composed of thin bands of a hard calcareous sandstone that alternated with thicker bands of an orange-brown rock that was much softer and more easily weathered. From afar this gave the whole cliff a quite dramatically striped appearance, with the stripes all sloping gently to the east along the dip of the outcrop. The layers of sandstone

---

an engraving of figures digging a canal, and turned out to be the millennial medal for contributions to geology, no less than the recently named William Smith Award. I took this coincidence as a most favorable augury for the rest of the journey.

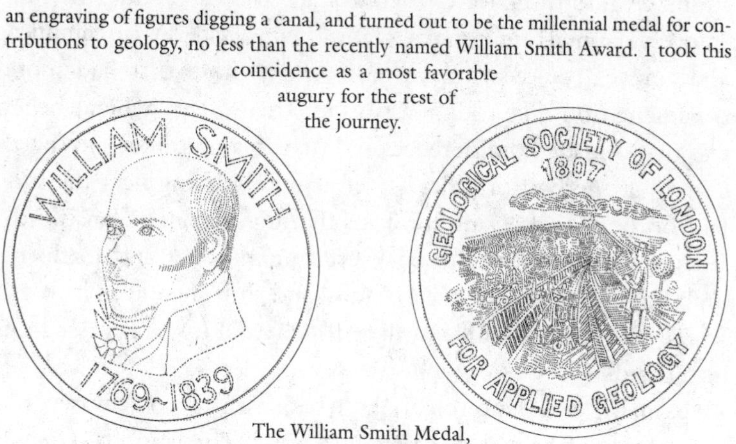

The William Smith Medal, showing on its reverse a scene from his canal excavations.

progressively thinned higher up on the cliff face—until, at the very top, there suddenly seemed to be a layer of a rather different rock, a thick band of a mustard-colored limestone that overhung the rest of the cliff and had evidently proved alarmingly dangerous to whoever farmed or walked or played golf on the meadowland at the top. Chunks of this rock lay littered below on the beach—showing where either human beings had deliberately knocked overhanging blocks of it down from the summit, or where erosion had caused it to tumble.

It was from within the shattered remains of a huge boulder of this particular rock that I had found my polished-looking ammonite all those years ago. Now, nearly half a century on, I looked for another. I looked long and hard, but after an hour or so admitted defeat. I couldn't find one anywhere—although I did find one or two fragments of very different-looking specimens, bits of ammonites that did not look polished at all, and which, to judge from the sandy limestone particles adhering to them, seemed to have fallen from some of the bands of rock lower down the cliff face.

In finding these I was reminded of the long debates we used to have in our Oxford paleontology classes, about whether smooth-skinned ammonites, with their decidedly streamlined appearance, could swim any faster than those that had more ornamentation and roughness about them. The general belief was that it made no difference, and that an ammonite—no speed demon at the best of times—lived its life bobbing near the surface of the sea and using the arms that hung from its backward-facing aperture merely to rock itself gently back and forth, to grab hold of passing morsels of food and stuff them into its jaws, where they were roughly dealt with, chewed up, and passed on into the digestive tract.

Nonetheless, the guidebooks I had with me on my second visit readily identified that first ammonite. The smooth-skinned cephalopod turned out to have been a fine example—how I wished now that I had managed to hold on to mine!—of *Leioceras opalinum*, a creature that defines by its very presence a junction point of immense importance within the English Jurassic.

The rocks that customarily lie immediately *below* what is called, eponymously, the *opalinum* zone, belong to the upper part of the epoch known as Lias, and which itself is chronologically a part of the lower part of the geological period known as the Jurassic. Then again, those rocks that occur (at least, occur in this part of the world) immediately *above* where this fossil is to be found, belong to a formation called the Inferior Oolite, and which lies within the epoch generally known as the Middle Jurassic.* *L. opalinum*, the small, smooth, reddish disk of the ammonite I had once held thus forms the boundary marker between the Lower and the Middle Jurassic, between the Lias and the Dogger.

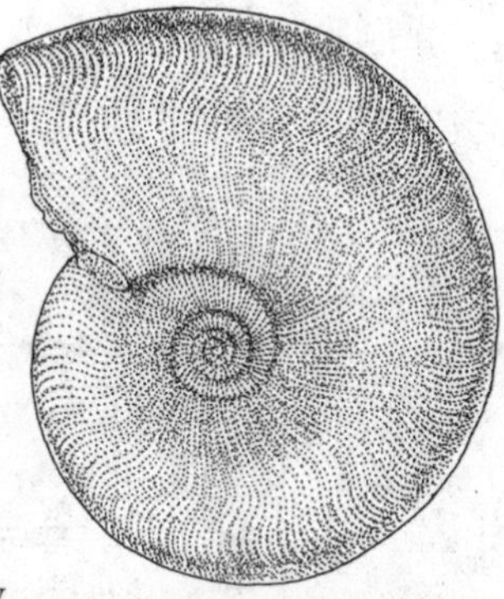

Leioceras opalinum, a defining smooth-skinned ammonite of the Lower Jurassic.

This smooth-skinned ammonite, the palm-handy stone I had once contemplated skimming across the surface of the Channel, records one long instant of faraway time. It was an instant that can be proved, by the radiochemical techniques that have so advanced the business of chronostratigraphy, to have passed more or less exactly 178 million years ago.

The fossil that I had once held in my almost-six-year-old hands was one of the most powerful keys that would unlock the

*The Middle Jurassic is known also as the Dogger epoch; and the Upper Jurassic, the Malm. The terms, more widely used in Europe and United States, are now internationally recognized, though this classification—all falling within the overlapping jurisdictions of academia and government—is subject to interminably long squabbles.

secrets of Jurassic time. *Lioceras opalinum*, indeed, was a fossil—one of very, very many in his collection—that William Smith, with his theory of fossil uniqueness, would come to know very well indeed. It was a great shame that I had lost it: I would have liked to have carried a specimen with me, as talisman, and try as I might, I could not find a replacement.

To have discovered one would have been ideal; and yet my memory of that first one remains comfortably powerful enough. As I walked back up the cliff to the parked car it occurred to me that my hero and I were now subtly connected by this single small lozenge of limestone and calcite, by the smooth and silkily beautiful physical object that he had once held in his hands, much as I had done all those years later. To Smith back then, and to countless other inquiring collectors since, the little fossil was very much more than simply a four-ounce reminder that tropical oceans had existed in the England of 178 million years ago. It was a symbol also of the beguiling magic and mystery of the science of modern geology, and provided a cozy link between the past and the present, between the extremes of the ultramodern and the ultra-ancient.

✣

The Carboniferous period in Britain, which began 360 million years ago, was generally a very lively time. It was an age characterized by torrid wet heat, by shallow seas and endless swamp, by thick rain forests crawling with slimy amphibians, a time of huge dragonflies and tall ferns, of warm deltas with pools that supported trilobites and shellfish. And then, with what can only be described as a fit of geological suddenness, everything changed, and all the life that was richly abundant dwindled to nothing, mysteriously withered away. In the Permian and Triassic periods, what is now the continent of Europe was dominated by endless sandy wastes, blasted by hot dry winds. Lifelessness, aridity and blistering heat suddenly took the place of all that Carboniferous moisture and fecundity.

There is an explanation for all this, an answer satisfactory to all who have wondered, and that has only been newly found, buried in the arcane complications of plate tectonics.

The early world was a terrifyingly volatile place. It was always mobile, its crustal blocks caught up in violent swirls of ferocious movement. It now seems clear that during the Permian one of these immense million-year-long swirls saw to it that all the protocontinents were briefly fused together—yet again; they had been fused many times before. In doing so this time they formed the now-familiar supermass known as Pangea, the continent that was the true beginning of earthly everything.

There were very few internal seas within Pangea: Such oceans as existed in Permian times lay principally at the peripheries of the giant landmass. Within, arid and windswept, were huge plains and mountain ranges and salt flats—either bitterly cold or raging with heat. Mongolia, the northern Chinese plains, the Bolivian *altiplano*, the Arizona desert—all of these look today as the corner of Pangea that would ultimately form Britain must have looked in Permian times: hot, dry, and very stable.

Except, with terrifying suddenness, another series of massive convulsions broke out. The oceans roared back in to inundate the plains. It must have been the almightiest of spectacles—much as if the South Pacific were to rise today, to flood all of China except its highest peaks, and then to lap hungrily away at and eventually flood to great depth the immensity of the Gobi Desert. And with the water came life, which burst forth once more and was soon teeming in wild profusion.

The warm oceans were benign and fertile again, and filled with living creatures—living marine beasts that, much more significantly for today's geologists, died. Over the millennia the skeletons of a thousand species of their occupants, ammonites to belemnites, brachiopods to oysters, ichthyosaurs, and plesiosaurs, rained down to lie in vast thicknesses on the ocean floor below. In the jungles there were dinosaurs, as well as scores of types of early birds and dozens of strange creatures that had evidently crawled up from the ocean and out of the freshwater ooze, and that had readily joined in the battles for territorial supremacy.

The record of that sudden new explosion of life remains today firmly locked in the limestones and shales and sandstones of the Jurassic. Fossils are everywhere in the thick masses of Jurassic strata, attesting to an abundance that is the very opposite of the lifeless tedium of the Permian. And no matter that the name *Jurassic* itself comes from the Jura Mountains of Central Europe: it is an unequivocal reality that the best expression of the outcrop of the type rocks of this period is in the long yellow swath that William Smith confidently painted running southwest to northeast, right across the center of England.*

The conditions were not exactly similar during the millions of years of England's Jurassic. The seas were shallow and warm only across the very center of England. Elsewhere the presence of huge landmasses made critical differences. Two of these landmasses above all help shape the outcrop that William Smith mapped—and across which I made my short pilgrimage.

The first, to the north, was a giant body of land called the North Sea Dome, which extended eastward to Russia and beyond, but in the west had spurs of high hills that ran across into Scotland and along the spine of what are now the Pennines. (No Jurassic rocks, of course, were laid down here, since no sea was present in which the sediment might be suspended.) To the south was the second, a low island fifty miles wide and oval shaped, extending over what is now the Brittany peninsula. The

---

*Uniquely in the world, the British Isles manage to display rocks from very nearly every single one of the geological epochs, from the oldest, the Precambrian and the Cambrian (from 1,000 to 570 million years ago) to those of the Pleistocene, which began as recently as 1.65 million years ago. Only the Miocene, which stretched from 20 to 5 million years ago, and during which the first hominidlike creatures appeared, is less than wholly represented. The remarkable appearance of well-nigh all ages and types of rock in the British Isles can tempt the belief that it is right and proper that the science of geology was born in Britain, and further adds emotional claim to William Smith being its most natural father.

coastal seas that lay directly off these two landmasses were, as one might suppose, muddier, less rich in oxygen and very much shallower than those further away—and the Jurassic rocks that were laid down in them, and in the estuaries and sand flats that abutted them, were much less likely to be the thick-bedded limestones characteristic of oceanic conditions, and much more likely to be thinner and alternating layers of clay-silt-sand-limestone, clay-silt-sand-limestone that are the inevitable consequences of the ebbs and flows of the oceanic edge.

Which is precisely the case in South Dorset. As I wound my way slowly northward from West Bay, the rocks that the fossil record insisted were of Middle Jurassic age were a thin and confused mess of sediments. Evidently they had been deposited in shallow seas and estuaries, in basins that had been subjected to frequent earthquakes and faults, uplifts and downthrusts. And in all cases the outcrops displayed the so-called rhythms of the cyclothems—with clays deposited where the waters had been deep; then bluish or yellowish sands, as the waters began to recede; then limestones, becoming progressively finer- and finer-grained as the waters reached the climax of the ebb, and deposition came to a virtual end. After which the waters came flooding back, and with them the clays, the silts, and, as the shallowing began all over again, the sands and the limestones once more. Cyclothems are a feature of edge-of-ocean sediments: Dorset has them in abundance.

Along the outcrop the villages were adorned with stunningly pretty names—Salwayash, Bradpole, Beaminster, Melplash, Haslebury Plucknett, Ryme Intrinseca, Chedington, East Coker, Mosterton, and Netherbury. The countryside here was intimate, small scale, constantly changing—the softer limestones had been eaten by the acid rains and formed valleys, there were copses of pines growing in the sandy soil where caps of sandstones overlay and protected mounds of shale, and roads passed deep into gorges of soft brown sand that came away if you scratched the walls with a fingernail.

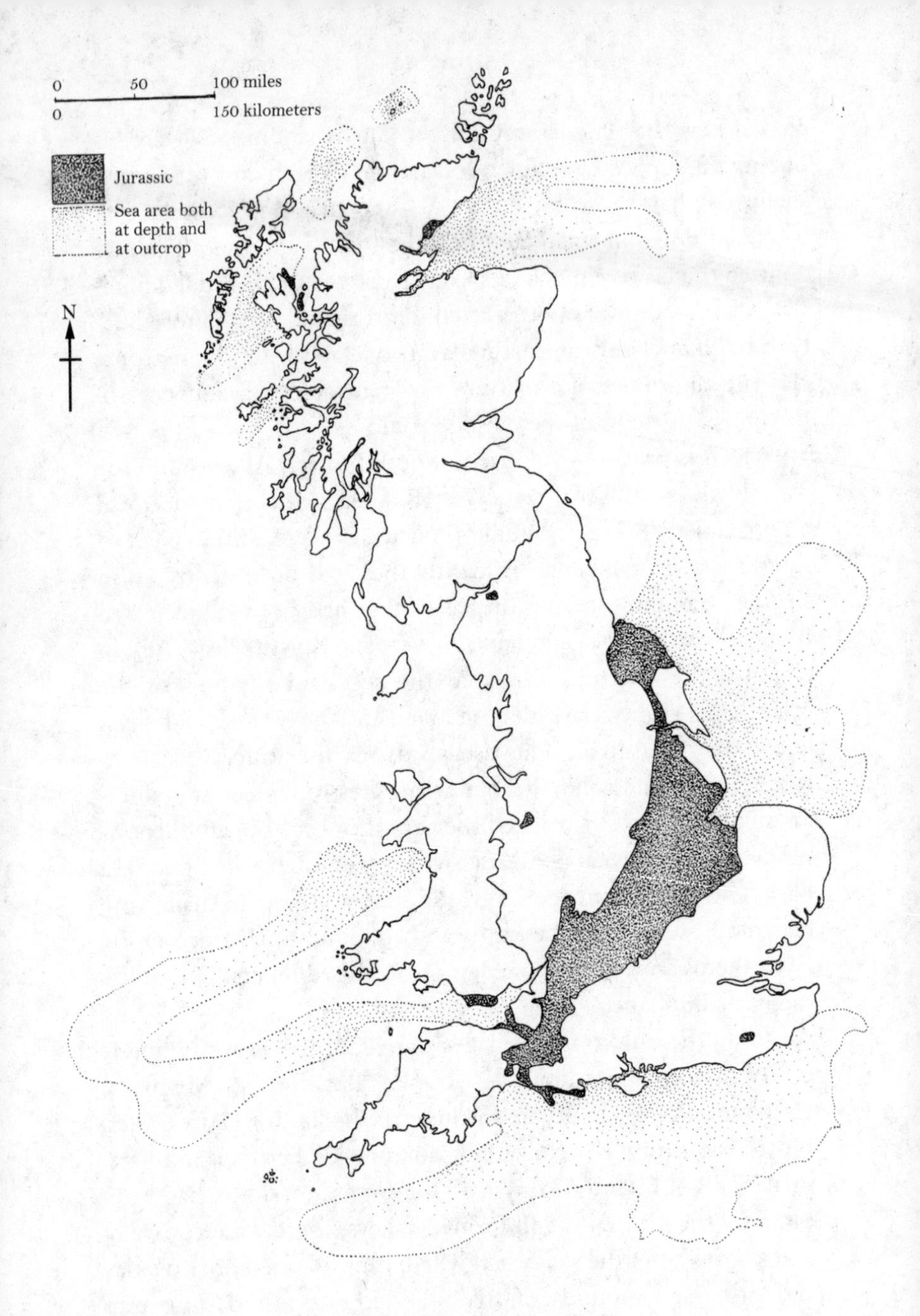

The main outcrop of the Jurassic rocks in England.

The Jurassic rocks here had been given names that were almost as sonorous as the villages on which they lay. The Bridport Sands is perhaps rather prosaic, the Yellow Conglomerate a little less so—but what of the Wild Beds, the Red Bed, the Snuffbox Limestone,* the Forest Marble, the Fuller's Earth, the White Sponge Limestone, the Scroff and the Zigzag Bed? And if not magical names, then magical appearance: What could be better than the famously extraordinary outcrop, found in a low cliff to the west of the old coastguard station at Langton Herring, which is ten feet thick and made up of nothing less than a solid mass of crushed and flattened specimens of the famous Jurassic oyster, *Ostrea acuminata*?

The village houses here—most of them ancient, many of them thatched, twined with roses, and huddled into cozy valleys—are generally built of limestone blocks that had been hauled from quarries some miles away, and which could not fairly be said to represent the mishmash of rocks below. This offers a sharp contrast to villages in the Cotswolds farther north, where the underlying Middle Jurassic is an inescapable feature of the architecture overhead. My grandparents had lived for a while in Symondsbury, a pretty village lying a mile or so to the west of my main Jurassic track: Their house was built of Portland stone, an elegant building stone (Saint Paul's Cathedral; the Ashmolean Museum; much of colonial Williamsburg) that actually comes from near the top of the Upper Jurassic, close to its junction with the Cretaceous. I remember seeing in the old house walls traces of the biggest of all ammonites, the famously lumbering *Titanites*, which is a marker fossil for the Upper Jurassic; someone also once pointed out part of the internal cast of the shell of another, odder beast: a small, long, conical, spiral-formed snail

*This is a Middle Inferior Oolite horizon characterized by the presence of countless small hard nodules of iron-rich concretion: The local quarriers thought they looked like snuffboxes, and called them so.

known as *Aptyxiella*, which vulgar quarrymen liked to call a Portland Screw.

To the north of Crewkerne, in Somerset, the Middle Jurassic rocks' lithologies begin to change. I was now leaving the rush and turbulence of the cyclothemic shore, and coming to a point where the seas—shallow but stable—are more oceanic than coastal, less disturbed, less turbid. The limestones become purer and thicker and more resistant—both more resistant than the rocks in Dorset that are of the same age and, more significantly, more resistant than the rocks in Somerset, Gloucestershire, and Oxfordshire that are younger, and lie on top of them. Because of this the outcrop of the Middle Jurassic everywhere north of the Dorset-Somerset boundary is marked by a sudden, steep hill: As I traveled farther and farther north, so this hill, and the very prominent edge where it rises up and away from the softer rocks below, becomes ever steeper, more and more evident, more and more obvious.

Small wonder that William Smith found the area around Bath the most congenial for his studies. Not only was it an attractive town, jammed with interesting personalities and lively minds: It was also happily sited at a place in the country's immense geological mosaic in which the Middle Jurassic rocks outcrop in a blindingly obvious way. The general line of their outcrop, which extends all the way north from Dorset to the Humber in Yorkshire, some two hundred miles, is one of the great dividing lines of world geology, once seen, never forgotten. Around Bath, close to where a northbound traveler like me today, Smith two centuries before, first comes across it, it is stupendously memorable.

On the western side of the line are the timid, milquetoast clays and weakling shales of the Lias, of the Lower Jurassic; on the eastern side are the tough, thick oolitic limestones of the Middle Jurassic. On the western side the consequential scenery all is valley and marsh, river course and water meadow, lowing cattle and in high summer, a sticky, sultry heat. On the eastern side, underpinned by the limestone, everything has changed: there is upland plain and moor, high hills, high wind and flocks of sheep, and in winters fine white snows blowing on what can seem an endless and treeless expanse.

And on the very line itself, at the point where England has tipped itself up gracefully to expose the limestones at its core and to reveal the huge physical contrast between their hardness and the silky softness of the Lias clays below, is a long, high range of hills and cliffs. The line is, for the most part, an escarpment edge that rolls far to the horizon, separating vales and downlands from high plains and uplands.

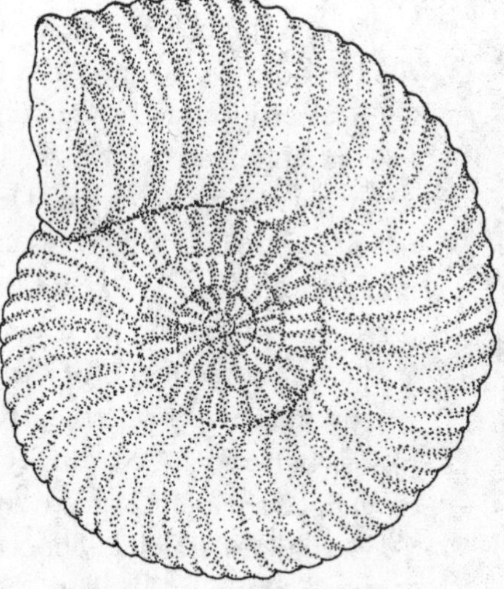

Titanites giganteus, the largest of the British Jurassic ammonites, can be found up to three feet in diameter.

We see this line in scores of places. Down at the southern end of the country—the Bath end—we see it where Crickley Hill and Birdlip Hill rise hundreds of feet above the town of Cheltenham. We see it where Wootton-under-Edge (a village set on Lias clay) nestles below the village of Oldbury-on-the-Hill (on Middle Jurassic limestone). We notice it, we *feel* it, when we drop sharply down from it via a dangerously twisting switchback road as we descend westward from the high plains of Snowshill (on the Middle Jurassic) to the antique shops of the clay-valley town of Broadway (on the Lower Jurassic). We can see it unroll over a dozen miles if we drive along the traffic-clogged roadway of the A46, on the stretch between Bath and Stroud: On going north, everything visible to the left is Lower Jurassic clay, that hunches low to the horizon: Everything to the right is Middle Jurassic limestone and rises high, its edge topped with oaks from which big black crows take in the view of the grassy fields below.

We see the phenomenon exhibit itself over and over as we rumble northward across the land—we see it through central Gloucestershire and Warwickshire, through Rutland and Leicestershire, across Nottingham and Lincolnshire—so that when, a day or two after I had left the warmth of Dorset, I found myself in the cold of Lincolnshire coasting along the A15 northbound from Lincoln (where there stands a fine Jurassic cathedral, made of just the same-age limestone as that at Wells, down at the far southern end of the outcrop) to Scunthorpe, almost exactly the same held true. To my right rose high limestone plains, buffeted by North Sea winds, dotted with sheep, flat enough and suitably exposed for the building of great Royal Air Force runways, training schools, and hangars. To my left, all was unsubtly different: A yellow cliff fell steeply away, and below it, spread almost flat like an unfolded survey map, lay a long, low valley, thick with farms, populated and cozy. The Middle Jurassic formed the upland landscape to my right; the Lower Jurassic the lowlands to my left.

Seen today, this pikestaff-plain exposition of how the rocks below make the hills above has the standing of a classic of English geology. Not a single guidebook to the geology of England fails to provide an illustration of the great hills at Crickley or Snowshill or the scarp face above the ancient wool town of Wootton-under-Edge. And, knowing now what William Smith had realized, explained, and then colored onto his great map when he saw this vast extent of hillside two centuries ago, "It all seems far too easy!" I find myself tempted to exclaim, as Thomas Huxley cried out on first reading Darwin: "How very stupid not to have thought of that before!" That William Smith was first to do so was a measure of the man's extraordinary achievement: to see what others could have seen but never did, to set down on paper what others might have suspected but never felt confident enough to declare.

The rock type that appears most consistently across the hundreds of miles of the Middle Jurassic outcrop is invariably an oolitic limestone. The term comes from the tiny, fish-egg-like concretions of which the limestone is made—and which, in some of the coarser-grained oolites, can easily be seen with the naked eye. Under the microscope the individual ooliths look like pearls, or mother-of-pearl accretions, or those many-layered gobstoppers English schoolboys once knew so well*—and yet in place of the crystal of aniseed at the heart, there is in the case of the Cotswold ooliths a tiny piece of quartz, or a fragment of shell. Layer upon layer of calcium carbonate had been deposited around this nucleus: Centuries of rubbing and jostling in Bahamas-warm blue seas then kept the slow-forming ooliths small and smoothly spherical; and their color, freshly formed, has invariably either been carbonate white, or else, if there happened to have been more than the usual complement of dissolved iron in the waters of the time, reddish brown.

But hundreds of years of gentle weathering have changed all of England's Jurassic oolites into the masons' most magical building stone. The physical nature of the rocks alone has long rendered them a sculptor's dream. Their bedding planes are straight and well defined, and either the chunks of quarried rock divide naturally into huge freestones, durable and massive and ideal for making big, strong, impressive buildings, or else can be sawed into flat sheets of ashlar, which can be bonded to rougher rocks and give, less expensively, the same harmony and integrity of look and color. Once in a while relics of the old coral reefs are to be found—for Jurassic waters were blue and shallow, much

*And known to their North American counterparts as jawbreakers.

like the Caribbean waters of today—and make for good building stone too: But they are knobby and ragged, the quarrymen call them hardstone, and most locals think of them as less pretty than the oolites themselves.

This "Cotswold stone" has a well-deserved worldwide reputation: The stones are solid, easy to work, more than amenable to simple carving, perfect for bearing large loads. Above all else, though, the Cotswold stone is quite simply beautiful. The humblest of workmen's cottages, if fashioned from a Cotswold oolite, is a lovely thing to see—and a huddle of the warm-looking Jurassic stone houses, clustered amicably in some river-carved notch in the meadows, can be so lustrously perfect, so quintessentially English that seeing it brings a catch to the throat.

In the villages that did well enough from the wool trade to evolve into something more than a simple confection of pretty houses, oolite stone was used to make just about everything—the columns, quoins, and mullions of the marketplace, the walls and facings of the cottages, the manor house steps and dovecotes, the fonts, the choir stalls, the bell towers and the transept floors of the parish church. In one Cotswold town, Bradford-on-Avon, there is a stone bridge with a stone jailhouse at its center point—durable, indestructible, and prisoner-proof.

The walls between fields, too, are built dry-stone, of oolite—and the contrast between a countryside that is apportioned by hedges and its neighbor with its fields divided by walls can be blisteringly sudden: In Lincolnshire, where the escarpment of the Lincoln Edge thrusts up with an immediacy reflecting the narrowness of the outcrop there, hedge becomes wall in just a matter of yards. Glance down from the edge, and every field is surrounded by blackthorn, hornbeam, and dog rose; glance up, and the fields are broken up by long stands of rough and mottled stone, tough, enduring and very, very old.

And yet it is the weathering, more than the simple durability, that provides the magic of the Cotswold oolites. The pale-honey coloration of newly cut stone, the rich orange and creamy reds of older, well-weathered buildings are quite sublime: The rock has given entire villages—Bourton-on-the-Water, Stow-on-the-Wold, Cirencester, Chipping Norton—an architectural wholeness; and it is a fine piece of irony that the poorer old homes, made of lesser-quality stone, have weathered in far more interesting ways than have the great houses built of the finest, unblemished freestone blocks. The patchiness of the weathered yeoman stone seems to give an extra dash of character, an additional element of charm.

The Middle Jurassic Great Oolite has given a distinctive look to entire cities too, a look that is now a part of the essence, the quiddity, of their very being. How would the crescents and circuses and terraces of Bath look if they had been constructed of some lesser stone? How much less lovely would Oxford be if the oolitic freestones of so many colleges and noble buildings had not been available, and the structures had been fashioned instead from a dour and ancient granite like the stones of Aberdeen, or from a grayish Millstone Grit of the North, of Silurian shales and gray slates from Bethesda, or made from a dull red sandstone like that found in unexceptional Middle English cities like Northampton, Doncaster, and Carlisle? The Middle Jurassic limestones make for a building stone of such gentle loveliness that its Cotswold walls, as J. B. Priestley wrote, know "the trick of keeping the lost sunlight of centuries glimmering upon them."

There is a little more complexity to the oolite than merely to say that the Middle Jurassic—even the outcrop between Bath and Lincolnshire—is composed entirely of it. There are, as William Smith himself found and described early in his career, at least two types of oolitic limestone represented in the epoch—an Inferior Oolite at the base of Middle Jurassic and a Great Oolite at the top of it. Between the two lies a layer some hundreds of feet thick of a distinctive claylike rock that, since it had the useful quality when mixed with water of *fulling*, or leaching the greasy lanolin from lambswool, was (as already noted) called Fuller's Earth.*

Within the oolitic horizons there are countless variations—of color (gray to buff, orange to ochre to pale scarlet) and fineness of texture, size of oolith, and width of banding and bedding plane. I chanced during my journey upon a roadside quarry near the village of Northleach, and the ebullient owner happily showed me around, pointing out with delighted pride the different colors and thicknesses of his rocks, and the uses he could make of the various types.

He used to be a farmer, but he had given it up back in the 1990s when he decided there was more money and amusement to be had from working the land below than trying to make ends meet from the land above. He had become a canny businessman: Build the Future, was his slogan. Use Natural Stone. He had handsome brochures, tempting customers with warnings—such as the charming thought that lichens would grow on untreated stones, that fragments of shell might protrude from walls and mantles, and that the presence of something called quarry sap might make the more delicate pieces crack in times of severe frost, and advising they be protected with hessian or burlap sacking.

He sold oolite by the ton for houses and walls; he also had a small gang of resident masons who slaved away in dusty warehouses making lintels and fireplaces, finials and corbels, carving gravestones and coats of arms. An admirer of English stone, currently living in Saudi Arabia, had recently ordered a globe, three feet in diameter: It would take the quarry owner's chief mason

*The active ingredient of Fuller's Earth is a so-called expansive clay mineral, montmorillonite, which is a phyllosilicate in that it is made up of parallel sheets of silica between which water and other minerals can be accommodated, so as to give it detergentlike qualities. The mineral is distinctively pink and named for the French town in which it is found in abundance. It occurs in rocks as a decay product of volcanic rocks—suggesting that the Middle Jurassic in which it is found was a time when older, igneous rocks were being worn down by the action of water.

three days to finish it, so long as his excavator could delve deep enough among the freestone blocks to find a chunk big enough and free enough of faults to allow for the extraction of a one-ton sphere.

What his quarry did not sell, though, was the one lithological expression of the Middle Jurassic that is too rare to market anymore, and that is what the quarrymen used to call "the pendle"—the so-called oolite "slate" of the Oxfordshire village of Stonesfield (not more than a couple of miles from Churchill, where William Smith was born). The rock is not a slate at all, but simply a very thin-bedded version of the usual limestone, and which has for centuries been used to finish the roofs of Cotswold houses. A truly classic manor house, that one might find between Oxford and Bath, should have its walls made of Jurassic freestone or faced with oolite ashlar, and its roof made of overlapping shingles of Stonesfield slate.*

The quarry sap mentioned in the Northleach brochure plays a vital part in the making of Stonesfield slate—which, not being a proper slate, cannot be manually split along its bedding planes. Quarrymen would only remove the pendle—they had to mine for this rock, using shafts and burrowing along narrow horizontal tunnels—between October and Christmas, and then leave the rocks set in clamps until the onset of the first sharp frost. The moment that this frost came—invariably in the middle of some January night, and the church bells would have to be rung to summon everyone—the miners would dash out to their waiting clamps of pendle and see if the iced-up quarry sap had done its work, shattering the limestone into tile-thin slabs.

---

*A very similar thin-bedded roofing limestone is to be found sixty miles farther north, in Rutland, where it is known as the Collyweston slate. But the Collyweston manifestly does *not* belong to the same horizon, despite its similarity: Its fossils (and here is William Smith's discovery being put to work) show it to be Inferior Oolite in age, whereas the presence in the Stonesfield Slate of the oyster *Ostrea acuminata* and the similar bivalve *Trigonia impresa*—as well as an impressive range of fossil vertebrates—show that it belongs to the Great Oolite, and is thus five million years younger.

If it had, then the men spent the summer cutting and shaping the slates—120,000 of them in one Victorian summer at the Kineton Thorns pit alone. But if it had not, then the uncracked slabs of pendle would have to be buried in the cold earth to preserve their sap—for if this quarry water was allowed to leach out it could never return, the rock could never be split, the slates would go unmade, and the men would go unpaid. It was a harsh and unforgiving business. Few are the industries that require frost to keep them in business: The villages of Stonesfield and Collyweston, unique in the land, came to rely for centuries on the regular onset of bitter cold, just as a Punjabi farmer would rely, year after year, on the coming of the monsoon.

The quarrymen would have another source of income, though. Fossils were to be found in huge abundance in the Stonesfield slate—and not just mollusks and belemnites, the standard fare of the collector and the academic. Here there were pterodactyls, dinosaurs, crocodiles, fish, cirripedes, annelids, starfish, and plants. Each time a quarryman or a slater might find a spectacular occupant he would pry it carefully from its limestone resting place and prop it tantalizingly in his cottage window. Geologists from Oxford University, on the hunt for specimens for teaching or research, would know to pass by Stonesfield each weekend, and scores of guineas would change hands for some of the choicer finds.

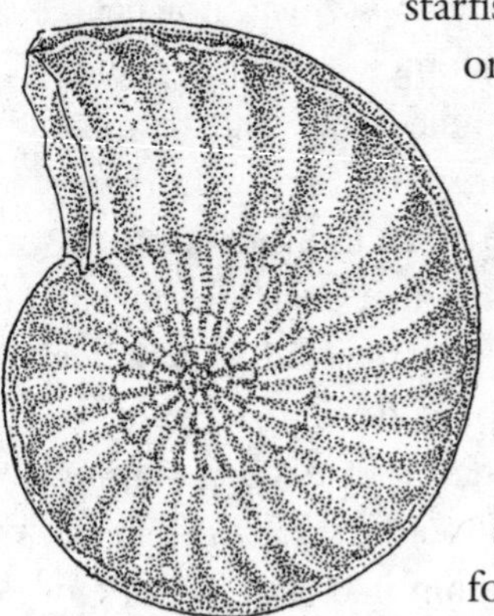

A classic Lias ammonite, *Asteroceras.*

After a pleasing week of wandering along the southern and eastern Midland outcrops of the oolites (and which were invariably garishly covered with the bright yellow flowers of *Brassica napus*, oilseed rape, which flourishes noticeably well on the limestones, with their six inches of highly alkaline topsoil), it was time to cross the Humber, and then watch the Middle Jurassic very nearly vanish. For some while, once I had arrived at Lincoln (where I bought for thirty pounds [about forty-five dollars] a nice varnished specimen of the Liassic ammonite *Asteroceras*) I had been aware of the looming presence to the east of the uplands of the Cretaceous chalk—which William Smith had recognized from forty miles away, that day in 1794 when he climbed up the tower of York Minster. Now the chalk pressed closer and closer, until, just outside the Yorkshire village of Market Weighton it lay not—as it should—on an Upper Jurassic horizon, like Portland Stone or Purbeck Beds, but on the Liassic rocks of the Lower Jurassic. At Market Weighton there were no Middle Jurassic outcrops at all: Northward up to the village of Thixendale the chalk lay directly either on the Lias or, for a very few miles, directly on top of the Permian.

An entire geological epoch was missing—and for no more complicated reason than that during Middle Jurassic times there was a lozenge-shaped axis of uplifted high ground to the north of Market Weighton, where no waters could lap and inundate, and where no rocks, limestones, shales, cyclothems, or marlstones could ever possibly accumulate. For half an hour I drove north through a series of villages that had never known the Middle Jurassic, knew nothing of oolites, had few fields of oilseed rape and that seem to have been held economic hostage to their geology after being surrounded on the one side by chalk and on the other by clay. The village houses were uniformly ugly and built of brick; the fields were divided, if at all, only by straggling and ill-kempt bushes. I felt strangely prejudiced against the place; and Smith's diaries showed he hurried through the countryside here too, eager to get back to a part of English geology for which he felt a keen affection but which here had found no place to settle.

North of Acklam, Leavening, and Burythorpe it returns once more, however—though this time in very different form. Just as with the looming presence of Yorkshire's chalk wolds to the east, so now another clump of hills began to rear darkly in front. These were the North Yorkshire Moors, heather-covered and wild, incised by deep valleys, and topped by vast flat acreages of unspoiled parkland.*

But the rocks here are not oolitic limestones—because the seas here were neither shallow nor Hockney blue. This was the delta of an immense river that coursed down from the North Sea Dome much as the Rhine courses out from Germany near Rotterdam. The rocks are thick—the river must have been Yangtze vast and sent billions of tons of sediment into the sea each of the five million years of its existence—but they are not limestones: They are sandstones and thick shales and, near Cleveland, ironstones. There are few marine fossils here, but a huge number of specimens of trees and fossil plants—horsetails, gingkos, cycads—and even on the Yorkshire coast, very thin bands of coal. This is the Middle Jurassic still: But as one can tell from the purple billows of heather and the cold and lonely roads snaking past secret radar stations and to the cliffs of the North Sea, it is made up of rocks very different from those at the other end of the country.

William Smith saw all of this countryside in the first two decades of the nineteenth century—he saw and traveled across and mapped all this and a great deal more besides. When I followed my own route two hundred years later I did so by dint of using a series of government-issued geological maps—most of them at a scale of ten miles to the inch, some at one mile to the inch, depending on where I was, where I wanted to be, and how much detail I needed to know. And in using those maps I would realize, each day that I pulled them from the glove compart-

*Spoiled in places, though, and usually by the military. The Americans have satellite stations at Fylingdales, and their soldiers listen from deep inside this outpost of the Jurassic, where they are surrounded by the fossil remnants of the earliest kinds of life, to the chatterings and secrets of our infinitely more modern creations.

ment, spread them out before me, and marveled at their accuracy and real beauty, that William Smith had, in essence, created them. He had been the first; it was his labors and his ideas that stood behind the artful confections that now guided me, and that guide a million others today.

There is something of a fine irony in the fact that my journey finished where it did, in this remote and enchanting part of North Yorkshire. For in the 1820s William Smith lived and worked here too, mapping and surveying an outcrop of the Middle Jurassic sandstones about five miles inland from the coast, at a village called Hackness. It was at Hackness that the fortunes that had eluded him for so many long and wretched years finally caught up with him, his life began to turn around, and he could begin afresh.

Now, however, with the Jurassic of the Dorset coast hundreds of miles behind, and with the Jurassic of Yorkshire before me slipping off the cliffs into the cold North Sea, is the moment to pick up the narrative again. To do so it is necessary to wind back the clock and return to 1802—to the year when William Smith, still young and energetic and with his later ill fortunes both unanticipated and a very long way off, embarked on what would become the most creative period of his life.

12

The Map That Changed the World

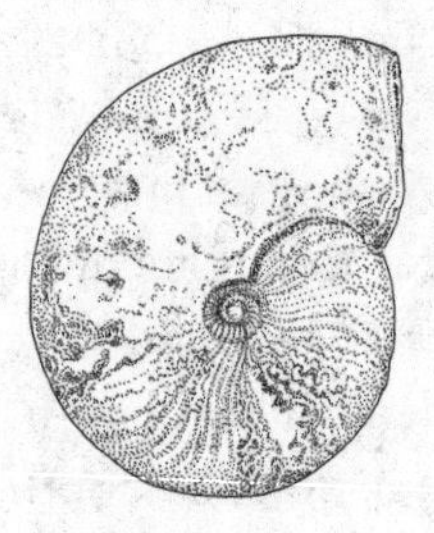
Clydoniceras discus

It was a work of genius, and at the same time a lonely and potentially soul-destroying project. It was the work of one man, with one idea, bent on the all-encompassing mission of making a geological map of England and Wales. It was unimaginably difficult, physically as well as intellectually. It required tens of thousands of miles of solitary travel, the close study of more than fifty thousand square miles of territory that extended from the tip of Devon to the borders of Scotland, from the Welsh Marches to the coast of Kent.

The task required patience, stoicism, the hide of an elephant, the strength of a thousand, and the stamina of an ox. It required a certain kind of vision, an uncanny ability to imagine a world possessed of an additional fourth dimension, a dimension that lurked beneath the purely visible surface phenomena of the length, breadth, and height of the countryside, and, because it had never been seen, was ignored by all customary cartography. To *see* such a hidden dimension, to imagine and extrapolate it from the little evidence that could be found, required almost a magician's mind—as geologists who are good at this sort of thing know only too well today.

And yet this was as yet a wholly unknown area of imaginative deduction—there were no teachers, no guidebooks. Just one man, doing it all by himself, imagining the unimaginable. Small wonder that the map of these new underground dimensions took fourteen long years to complete, years more than it was supposed to. It proved to be the financial ruin—at least in the short term—of the man who had the vision to see it made. But when it was done it all proved to be so very good, so revolutionary, so filled with potential for profit and fame that it was stolen, copied, pirated, and the man who had made it overlooked, ignored, and forgotten for years.

When the project was begun, at the very start of the nineteenth century, it was just like so many other vast and ambitious schemes of the time, instilled at first with sunny optimism and good humor. There was an absolute certainty that it would be done, and done very well. And it would indeed be done and done well—but, as it turned out, at what a cost!

William Smith, who would come to be known all across the country as the strangely driven and ever-wandering maker of this map, had ended the eighteenth century with a reputation for scholarship and *brio*. By the time the nineteenth century was properly under way, he was fast becoming something rather different: an all-too-familiar, hale, and friendly figure in the post-houses and inns along the English stagecoach routes. He was a very talkative man, with all the enthusiasm of the slightly dotty. He chattered so much about the passion for landscape and rocks that drove him that he soon picked up a nickname—"Strata" Smith—from the barmaids and fellow passengers and those who dined beside him.

And though to some he might have been a bit of a bore, he was by most accounts well liked—a bluff and hearty, muscular-looking man, full of energy, restless, talkative, jumpy, untidy—

and invariably to be spotted scurrying hither and yon, always scurrying, and carrying great bundles of papers, maps, and charts under his arm.

Maps, indeed, were an abiding passion. Like many uncontrollable passions, they proved a burden too. Money was a constant trial. When Smith was out working he had at least forty maps on the go at once, all of the largest scale and greatest accuracy he could find. And he found his need to have them dismayingly expensive. Because of the time-consuming way maps were engraved in the early nineteenth century, they were much more costly in relative terms than they are today—and Smith's purchases added significantly to what he was coming to realize was the very high cost of his own hugely ambitious cartographic enterprise.

More trivially he was to write that the simple business of carrying maps as he bought them, rolled up in tubes, made the practical side of his work very difficult. He was always spotting interesting pieces of rock beside the roadway—always jumping down from the driving seat of the chaise and clambering back up to sit beside the coachman (he had this childlike wish to sit up there in the wind, no matter the weather, the better to see the landscape). But the rolls of maps made this near-impossible: To make matters simpler he carefully cut the larger maps into small squares, bound them into a green morocco carrying case, and then was able to write in his diary much later, and quite smugly, that even after travelling thousands of miles with them the maps were "in pretty good preservation, though disfigured in some parts by speculative attempts at the delineation of strata ranging through parts I had not then seen."

He must have cut an eccentric figure. People were said to have been frequently rather frightened by him—he was enormously strong looking, and wore a rather stern expression, such that strangers would ask him if he had ever been a boxer or a soldier. In his journal he quotes a man remarking to him that he was "so damned good built" that it was likely he must have been a great walker in his time—which, given Smith's passion for strolling, clambering, marching, pacing and trolling across countless meadows in search of outcrops, was perhaps only the half of it.

He kept himself dressed ready for the moor, in brown tweeds and with a remarkable broad-brimmed hat (to keep off the sun). This, he wrote later, often prompted Dissenters—those religious nonconformists and free-thinkers who were flourishing in the fast-changing, fast-industrializing world of the time*—to stare at him, thinking he might be one of them (which, given that he was practicing a science dealing in large part with the evidence for and against Divine Creation, he was most certainly bound to be). The Quakers in particular liked to assume that "from a resemblance in habiliment" he was a member of their church, and they would wave as he passed.

Smith's motive for making a proper geological map of England was more complex than it seems. On one level he was driven by simple intellectual passion. He had discovered in Somerset that the rocks were spread out in historical order, and that their fossils allowed their underground arrangements to be delineated and predicted—and the corollary of these findings was quite obvious to him: He should make a map. He should make it if for no other reason than that his discoveries now meant *that he could*. But there were other reasons too, and these he spelled out from time to time in his writings: He had a deep, obsessively felt need to be given wider recognition for what he was sure were profoundly important discoveries.

---

*By dressing this way and dealing with the more landed of his clients, Smith was risking much: Dissenters were notoriously ill regarded by the older conservatives. Sydney Smith, the essayist and wit whose writings probably most accurately catch the mood of his namesake's time, wrote in 1807 that "when a country squire hears of an ape, his first feeling is to give it nuts and apples; when he hears of a Dissenter his immediate impulse is to commit it to County Jail, to shave its head, to alter its customary food, and to have it privately whipped."

By the time he ran into the difficulties with John Debrett, and was forced to abandon the book he had expected Debrett to publish, he was puzzled to find that this recognition was proving to be more elusive than he expected. That the general public was less interested in him and his ideas than he thought they should be, bothered him deeply. He wanted to be immortalized: He would become so by creating a truly impressive piece of work, one that would last as a memorial for generations.

And there was a financial incentive, too. Landowners, among whom his reputation was spreading like bindweed, were set to wondering whether there might be mineral deposits on their estates: Smith, who men like the duke of Bedford were saying had an uncanny way of predicting what kinds of rocks might lie where, might hold the key to the gentry's further amassment of wealth. Lord Egremont wrote to Smith to inquire if coal was likely to be found on his land at Spofforth; Thomas Johnes, a landowner and Member of Parliament in mid-Wales, offered a reward of five hundred pounds if he could show where he might find lime on his lands; and others deluged him with letters asking if he suspected there might be lead, or tin, flints, clay, marl, or sand.

And then there was the newly formed Society of Arts, which took the long-term, national view. In 1802, at the time when Smith's efforts were getting into high gear, William Shipley's "Society for the Encouragement of Arts, Manufactures & Commerce," which, as its name suggests, had been founded to support all manner of worthy schemes, offered a fifty-guinea bounty for anyone who might make "a mineralogical map of either England, Scotland or Wales." The society's leadership knew now that such a thing could be done: It was up to someone—someone with time, energy, knowledge, and, though it was never openly stated, sufficient funds—to take up the challenge and satisfy a demand that was now clearly swelling up from the country's body politic. England wanted a geological map. Rivalry was in the air. The country's cartographers were in the traces. A race was under way.

The society's challenge prompted Smith to lever himself out of his base in the countryside—where, he reasoned, he would win relatively few commercial commissions—and set himself up in an office in town. Accordingly, on April 3, 1802, he placed an advertisement in the *Bath Chronicle*, announcing his partnership with Jeremiah Cruse, one of the men who had subscribed to his proposed Debrett book, and the subsequent formal opening of the firm of Smith & Cruse, Land Surveyors:

> Having opened an office at No.3, Trim Bridge, the corner of Trim-Street, Bath, for the purpose of carrying on the above business in all its branches, we take this opportunity of informing the Nobility, Gentry and Public in general that attendance will be regularly given at the above Office, where orders will be thankfully received, and executed with accuracy, neatness and punctuality.

He brought his (by now enormous) collection of fossils up from Tucking Mill House and arranged them in their stratigraphical order, at first in boxes on the floor, then on a set of specially constructed sloping shelves. The duke of Bedford came to see them, and so did the duke's land steward, John Farey. And although Messrs. Smith & Cruse was not to be an entirely successful firm—Smith resigned from the partnership early in 1804—its existence consolidated his friendship with Farey, who was to become one of the most conscientious supporters, both of Smith himself and of the map he was set upon making.

In John Farey he could not have had a sturdier ally—although the relationship got off to a rocky start. Farey's initial career as the duke's land steward had been hatched mainly because he had been born on the Woburn estate, where his father had a tenant farm. But the young man's interests were in fact wider and more catholic by far than those of a typical ducal employee: He soon left Woburn (sacked by an incoming duke) and became an expert musician (and a chorister of note), a mathematician whose work (on the curious properties of vulgar fractions) is still known today, and a contributor to encyclopedias on such topics as astronomy, engineering, the history of pacifism, the design of steam engines, the decimalization of currencies, and the population theories of Thomas Malthus.

He was also hugely interested in and stimulated by Smith, and traveled with him frequently as a devoted acolyte and apprentice in those early years of the century, learning theories and techniques that he was eventually to put to good use on his own account. Rather too good, Smith was eventually to complain bitterly—in an incident that illustrates the growing problems, some real, others merely the consequence of his perception, that were beginning to cloud Smith's life.

The falling-out between Smith and Farey began when the editors of Abraham Rees's *New Cyclopaedia* turned to the highly literate Farey, and not Smith, when they wanted articles to illustrate words and concepts of geology such as *clay strata*, *coal*, *colliery*, and *extraneous fossils*. The choice greatly injured Smith's self-esteem, even though Farey was at pains to make generous mention in each article of his mentor's name, his achievements, and his theories.

But when Smith found out that the entry on *canal* had been cut back by the editors so as to exclude the entire account of his contribution to the building of the Somerset Coal Canal, he was cut to the quick. And then again, when, shortly afterward, Farey wrote to say that he had been asked by the national Board of Agriculture to prepare a survey of the county of Derbyshire, and was writing to ask whether, to save time, he might see Smith's own drawings of the strata so that he could trace them onto his own map, Smith became incandescent with rage.

No, he fumed, Farey most certainly could not see the map! If he wanted to survey Derbyshire, then he could do it himself, no matter how long it might take. And in any case, his letter continued, he, Smith, should have been the one asked by the board—not this "scientific pilferer," as John Farey, now a turncoat, had in his eyes become.

It was a brief-enough tiff—the men were the best of friends once again soon after, with Farey ending up as the most loyal of Smith's many supporters. But at the same time the argument shows once again the degree to which Smith felt he was being overlooked and shunned by those in society who were reckoned to be more gentlemanly than himself. The Board of Agriculture, for example, was composed of grand landowning men; their president, Sir John Sinclair, though ostensibly a friend of Smith's, was an indefatigable Scottish aristocrat, a man of vast wealth, who had raised his own regiment (the Rothesay and Caithness Fencibles), had become a world-class expert on sheep, and was interested enough in numbers to introduce the word *statistics* into the English language.

That so mighty a figure should apparently overlook in his choice for a county survey a member of the Oxfordshire peasantry, and give the task instead to an apparently learned man working on a duke's estate, was perhaps not all that surprising in the class-ridden days of the early nineteenth century. But to Smith it was a considerable slight, of a kind to which he would never become accustomed.

In the case of his relations with John Farey, though, the situation was later helped when Farey, too, was to feel the crushing weight of English snobbery—most notably being denied membership of the newly formed Geological Society, because he was merely the son of a farmer. In later years both Farey and Smith clung together in their mutually held opposition to what they saw as the lofty *hauteur* of the gentlemen rock collectors. So close was their relationship in those early days that Farey came to be regarded by friends as Smith's Boswell—a chronicler and lifelong advocate of all his master's works.

It was at about this time, too, that Farey introduced to

William Smith the one figure of great national repute who was going to be singularly important in the making of the great map—and the man who was to figure prominently in this growing dispute between the practical and the gentlemanly students of earth science. For Smith was not alone in regarding himself as a victim caught in the crossfire of a British class warfare: The entire and newly fashionable discipline of geology was about to be riven by arguments between the horny-handed toilers in the fields and the more fragrant dilettante practitioners, between amateurs who specialized in studying the land below and those regarded by the former as worthless dandies whose interest was more in owning and exploiting the land above.

The man who would try to mediate this dispute, and who would at first do his level best to accelerate Smith's progress in making his new map, was the then president of London's distinguished Royal Society, the influential and farseeing botanist, Sir Joseph Banks.

Banks—after whom an Australian flower is named, and whose reputation in New South Wales, of which he was a principal founder, is huge—is perhaps best known for being the moving force behind the notorious *Bounty* expedition of 1789,* which culminated in Fletcher Christian's mutiny against Captain William Bligh, and his subsequent establishment of a settlement on Pitcairn Island. He was not known to be especially interested in geology (other than being remarkably impressed by the columnar basalts of Fingal's Cave on the Hebridean islet of Staffa)—until he came into the ownership of a Derbyshire estate, Overton, near Ashover, that had valuable lead deposits.

But the lead mines were running out—Overton's famous Gregory Mine, which had once turned an annual £100,000 profit was now suffering losses of £23,000 a year. When Sir

*Banks established the expedition to ship breadfruit plants from Tahiti to the British possessions in the Caribbean, to use as cheap food for estate workers. He also first brought the mango to Britain, from Bengal.

Joseph borrowed John Farey from Woburn in 1797 to help drain his Derbyshire land, he mentioned his problem with the lead. At first there was nothing Farey could do—he was not very interested in geology and didn't know anyone who was.

But then he met William Smith at that great agricultural gathering known as the Woburn sheepshearing. He learned about Smith's planned book outlining the new principles of stratigraphy, which at that point John Debrett still seemed keen to publish. Smith, he felt sure, would be just the right person, and he explained this to Banks: If anyone could predict whether more lead might be discovered at Overton, it would be William Smith.

Banks was delighted with the encounter, which appears to have first come about in the summer of 1801—that, at least, was when it is first known (from John Phillips's later biography) that Sir Joseph "favoured Smith with an interview, and from this time until his death [in 1820] remained a steady friend and liberal patron of his labours." Farey encouraged Banks to help: In February 1802, after taking an expedition with Smith and another of Smith's pupils, named Benjamin Bevan, he wrote a long letter to Overton, insisting that Smith's findings were of paramount importance.

He explained to Sir Joseph, patiently but from the tone of his letter not at all condescendingly, that he believed Smith had made two great advances: He was able to document the *sequential* order of British rocks, and he was able to *identify* the rocks that made up the sequences, by examining the fossils found within them. Smith, to put it more simply, could both tell *where* rocks were, and *what* they were: And, by dint of an energetic and systematic program of observing, surveying, measuring, and marking, he could draw—as indeed he was now in the middle of drawing—a proper, detailed, and accurate geological map of the nation.

Sir Joseph took the bait. Smith first noted the likelihood of winning the great man's support—support that would prove invaluable, if it came—when he wrote, in May 1802, to a friend from his office in Bath:

> I have been obliged to have recourse to an uninterrupted pursuit of my subject, with all my plans and papers together here, and I shall call on Sir. J. Banks in London to settle about bringing my papers before the Publick in much better form than if they had appeared last year. I am now confident they are correct, and my map begins to be a very interesting History of the Country.

By the following month a new map—not much better than the first small-scale attempt he had made in 1801, presumably—was beginning to take shape, and he felt confident enough to show a version of it to Banks at the Woburn agricultural meeting. The *Agricultural Magazine* for the month reports the fact dryly: "Smith . . . exhibited his map, now in very considerable forwardness, of the strata of different earths, stones, coal & which constitute the soil of this island. He was particularly noticed by Sir. J. Banks."

The following October he wrote to Banks at his country house in Lincolnshire to bring him up-to-date, and then again in spring (all according to his journal, which he was now writing at a furious rate) he took the latest version up to London, to show it off to Banks in the Royal Society suite at Somerset House, off the Strand. The encounter was a small triumph of assurance and optimism. Banks would support a subscription for the new publication, and wrote a check for fifty pounds to get the project off the ground.

As he left his great patron's chambers, Smith must have felt that he now had every reason to believe he had won the support and the enthusiastic backing of the most powerful figure in British science: If he had until this moment from time to time felt overlooked and wronged, then at this juncture in his career he had precious little cause to do so. Later on, when Banks became increasingly exasperated with him, he might have some cause, but not now. Just now he could savor the sweet smell of impending success.

As it happened, though, the Somerset House meeting, which took place on April 21, 1803, turned out to be rather less propitious than Smith's delighted diary entry might make it seem—because it laid Smith open to a temptation to which he should not have succumbed: It seduced him into taking an apartment in London.

It was a woefully imprudent decision. Smith was not a well-off man. It had been nearly ten years since he had been in full-time employment, and although he was being paid handsomely enough for his work as a drainer or mineral surveyor, the work was typical of the freelance trade: days of feast followed by weeks of famine. And in between the episodes of paid work, he had to continue his journeying to make his endless geological surveys—journeying by stagecoach (for a not inconsiderable fee), lodging at coaching inns (for substantial sums), wolfing down pigeon pie and porter and the big breakfasts that his energies demanded. He asked the Society of Arts if they might help him out with an occasional supply of funds: They apparently turned him down, as did all too many others whom he approached. And now he was embarking on a program to spend even more.

He had an office back in Bath to run; a partner (the long-forgotten Jeremiah Cruse) to pay; and an estate at Tucking Mill to keep, with its stable and garden and a manservant on constant hand. On the face of it Smith was heading into deep financial waters—and yet at precisely the time when he should have been trying to keep his expenses down, he was suddenly seduced, presumably by the magnificence of Sir Joseph's own grace-and-favor* lodgings off the Strand, to take with effect from that same April day a permanent and costly lodging place just around the

*A Crown-owned property offered by the monarch as a lifetime residence to an individual who has performed great service to the nation generally or to the Crown particularly.

corner, at number 16, Charing Cross. He rented rooms from a Mr. Tapster—the building was called Tapster's Baths. He shared it with his landlord and with one Francis Place, a mysterious man described in the *DNB* as a "tailor and radical reformer."

The accommodation, though expensive, was small: Smith spent most of his London days at the Craven Coffee House close by, which he used almost as an office. His initial plan was to engage an assistant who could stay in the Tapster's Baths rooms during the day, and begin work on designing the final version of his map, as well as starting a program for drawing all his fossils. But for the time being he decided to leave his actual collection behind in Bath, at the offices on Trim Bridge.

It turned out to be one of the very few prudent decisions he was to make during this frantic period in his life. It was just as well that he never brought his fossils to town—for just a few months after he had moved in, and as a harbinger of other disasters to come, his building and the one next to it caught fire and burned to the ground. Smith's assistant, who would almost certainly not have been able to carry hundreds of pounds of boxes of precious samples out of a blazing house, managed to save most of his papers. They were removed "in a hurry and in disorder," according to John Phillips, but they were at least safe. Smith was in any case not there: Once again he had been afflicted by his persistent wanderlust, and when his house burned down he was out of the city, on yet another mapmaking field excursion.

The fire did not deter him from his grand plan to have an address in London: quite the reverse. He now took an even larger establishment—an entire five-storied mansion at number 15 Buckingham Street, in the Adelphi, an impeccably sited house standing beside the Thames, beautifully designed, conveniently located both for the Society of Arts and for Sir Joseph Banks's office at the Royal Society. He paid a rent of eighty guineas a year—far too much for a man with such slender resources, with rent to pay in Trim Bridge, a mortgage to pay on Tucking Mill, and with a wildly varying income—and added to his bills a further pound a month for a housekeeper, the charmingly named Mrs. Kitten.

A house in the Adelphi development was indeed a wonderful place for a man of means to live. The Adam brothers had designed them, and imported Scottish laborers, kept content by having bagpipes played to them, had built them. The houses, which went up on the site of the old riverside mansion of Durham House (immediately upriver from Somerset House) were spacious and elegant, and known by the clever honeysuckle motif on their stucco frontings. But the development was not a success, and the brothers only avoided bankruptcy by selling them off at bargain prices in a lottery. Not that Smith had the money to buy: For all his sojourn in Buckingham Street, he was obliged to rent.

Smith's principal home in London's Adelphi, at number 15 Buckingham Street.

Smith liked the place enormously—so much so that in late 1804, after the duke of Bedford had made an inspection tour of his fossils down in Bath, he had them all moved up to London—thousands of specimens, wrapped in paper or in small packets, and by now displaying no fewer than 720 different species. As with the scholar whose home is considered to be wherever his library stood, so it was with William Smith and his collection: This is where his fossils would lodge from now on; it was to be his official residence, a suitably grandiose symbol of the work in which he was now engaged. "I am happy to inform you that my fossils are now safely arrived in London," he wrote to a friend in South Wales in June 1805, "and are now arranged in the same order as they lay in the earth."

He liked the notion—erroneous, as it happens—that Peter the Great of Russia had stayed at number 15 when making his celebrated tour of England in 1698. And he would have thought it eminently suitable that the young Charles Dickens lodged there in 1833, by which time it had been converted into flats. But when he was there alone—except for Mrs. Kitten—he sported none of the trappings suitable either for a Romanoff emperor or a writer of great fame. Inside the house everything was modest to the point of monasticism. He was becoming fearful, evidently, of the possible financial consequences of yet more extravagance. In later, wiser times when he experimented with an autobiography he was to write:

> In London a tax-gatherer of one denomination or other is never long absent from your door. With their heavy hands my old rusty knocker too often made my high old house echo to the attics. I might have reduced taxes by stopping up windows and, indeed, by shutting up useless rooms. But it was only a house of call for me on my way through London, and a depot for my fossils; for I had no time to devote to the economy and comforts of housekeeping. I never half furnished it, never had a dining table; no carpets crossed my old oak floors, no rich curtains darkened my windows; and though my rent was high I had no expensive living, no dinner parties, no wine merchants to pay.

He had indeed no time to devote to the comforts of housekeeping. His travel diary for the years between taking up residence in Buckingham Street in 1804 and in 1812 making his first real breakthrough in the publication of his map, reads like that of a Fury.

His business as a drainer kept him traveling incessantly between Norfolk and Kent, between outposts in Wales and northern Yorkshire. The owners of the Hickling Marshes, for instance, had him stopping inundations from the sea between Yarmouth and Happisburgh. He was asked to go to Dolymelynllwyn, near Dolgelly in North Wales, to look at the slate quarries and inspect the soundness of a new embankment. He went to the top of Snowdon, the highest mountain south of the Scottish border; he examined the copper mines near Llanberis.

We find his diary telling us he is variously in Yorkshire, Lancashire, Somerset, Gloucester, Devon, Rutland, Nottingham. Here he is opening a coal mine at Torbock, near Liverpool, here he is directing a trial boring at Spofforth, there he is off looking at the cliffs of Witton Fell in Yorkshire. He examined coal outcrops at Newent, at Nailsea, and in the Forest of Dean; he built a series of sand-dune-mimicking embankments, all shells and marram grass, to help protect the South Welsh coast from flooding at Laugharne. He worked on planning another big canal, the Ouse Navigation in Sussex; he was called in to help because the Kennet & Avon Canal, in his old stamping ground near Bath, had begun to leak. A landowner in Buckinghamshire tried to find coal on his land and called in Smith, who put down some bores and told him he was wasting his time.

Scores of other would-be coal millionaires demanded his time, which he willingly gave for his customary two or three daily guineas, if only to prove the pointlessness of their hopes. He was persuaded, despite offering the advice that owners were squandering their money, to conduct a survey in Herefordshire, another near Wincanton, and a third within sight of the towers of the Oxford colleges, at Bagley Wood—knowing on each occasion, as no other person in Britain could possibly know, that there was as much chance of discovering coal as there was of striking gold. It came as a pleasant diversion when he was asked to perform a task that would not end in disappointment—as when he planned a series of improvements to the harbor at Kidwelly, a little Welsh port well known to later readers of Dylan Thomas,* or when he completed work on the Minsmere Drainage Scheme in deepest Sussex.

Occasionally there were more amusing or rewarding tasks, which brought him more than his simple *per diem* fee. In 1810 the Bath Corporation called him in because of an unparalleled calamity: The hot springs, which had provided the town with a *raison d'être* since before Roman times, suddenly failed, and no one knew why. (No one knew why there were hot springs in Bath anyway, considering the city's location on top of thousands of feet of congenially stable sediments, with not a volcanic fissure in sight.)

He rushed from London to the Bath Pump Room, examined the situation, and declared that the only way he could find out the reason for the failure was to dig a bore into the very spring itself. The city fathers were appalled: Never had an excavation been performed in these most hallowed buildings. But in the end, faced with the prospect of a well gone dry, they reluctantly agreed—and William Smith and a gang of navvies hired by the day burrowed down through the familiar limestones and clays of the Middle Jurassic until, after suffering in temperatures of 119 degrees Fahrenheit and melting all the candles they had used to light their way, they found the problem.

It was all the fault of the large bone of a great ox—or, as the official city report of the time put it, "some large ruminant." Somehow the bone had fallen into the spring, had become crystallized with pyrite and flint, and had rolled itself into the channel and blocked it. The waters promptly made another channel

*Dylan Thomas is buried at Laugharne, where Smith built the embankments: One of his finer early poems was *The Map of Love.*

for themselves, as waters do, and flowed out into the Avon somewhere else.

A few of the more suspicious members of the corporation said it was all really the fault of a new coal mine that was being dug at the time three miles away, at Batheaston; and there was a very angry movement to have this mine stopped up, to protect the integrity of the springs. But Smith went to Batheaston to have a look and decided that the two were not connected in any way. Bath's hot springs had failed because of a pyritized ox bone, and nothing else. He removed it, the hot and healing (and vile-tasting) waters began to flow again with greater vigor than before, and Smith became, just as he liked, the hero of the hour. And for good measure, he plugged the hole in the coal mine as well.

Much the same happened when he drained the infamous Prisley Bog, on the duke of Bedford's estate at Woburn. This he managed with such speed and ingenuity—and, moreover, published a brief monograph in 1806 on how he had done it*—that the Society of Arts awarded him a medal. He was naturally delighted—except that it gave him good reason to remember, since this was the same Society that was offering a premium for making the great map, that the years were ticking on, and still nothing had been published.

Sir Joseph Banks was also beginning to wonder why his golden boy had not delivered. It was now five years since the men had first met, and since the day John Farey had told Sir Joseph—with some prescience, seen in today's light—what a stellar figure Smith was destined to be. Banks had already contributed the sizable but quite easily affordable fifty pounds. He had persuaded

* *Observations on the Utility, Form and Management of Water Meadows, and the draining and irrigating of peat bogs, with an account of Prisley Bog, and other extraordinary improvements, conducted for His Grace the Duke of Bedford, Thomas William Coke MP and others; by William Smith, engineer and mineralogist*, printed in Norwich and published in London by Longman and distributed by booksellers in 1806, with 121 pages and two folding plates, can fairly be considered Smith's first major published work. It made precious little money for anyone.

dozens of his friends and acquaintances to subscribe to the impending publication. He had seen Smith at least three further times—at the 1804 Woburn sheepshearing, at the Smithfield Cattle show six months later, and in London early in January 1805. There had been much talk of the map, "soon to be exhibited for the information of the curious" according to an optimistic publisher's notice in the newspapers, in Banks's very own London library.

But since then there had been nothing, and influential men were beginning to grow impatient and exasperated. Smith had been expected both to prepare a great map, and to continue work on the treatise on strata that Debrett had been hoping to publish. But neither book nor map showed the slightest sign of being readied for an appearance, even in rough draft. Richard Crawshay, the Welsh landowner who had been an early supporter and who had promised money, wrote sharply about this to Smith in February 1806: "I am sorry to find that your promise of and my reliance on you for a publication of great importance is totally vanish'd. You will excuse my interfering any further in your affairs. I wish you well." It was a considered and stunning insult. Smith wrote a spluttering explanation, but Crawshay never replied.

Even the tolerant old Sir Joseph was worried. He wrote Farey to the effect that he "[did] not feel himself so interested in encouraging any new work" of Smith's. Other potential subscribers were holding back, worried that Smith might never complete his work. They knew that he traveled widely, that he worked furiously, that he constantly promised visible progress: But nothing emerged—nothing from the Adelphi, nothing from Trim Bridge, nothing from Tucking Mill House.

It was, for Smith, a terrible time. His book on draining had brought in no money. His most loyal supporter, John Farey, had been appointed, rather than Smith, to make an official survey of Derbyshire. It seemed suddenly as though there might be a conspiracy waged against him, a sudden drying-up of commissions, the beginnings of a muttering campaign. He had fallen out with the assistant at Buckingham Street, a man named Roope, whom Smith now described as "a book-learned, paragraph-writing coxcomb"—a remark that was about as rude as one could be in the London of the day without risking a duel.

Then there had been an innovative suggestion from Sir John Sinclair, the president of the Board of Agriculture, that Smith might become associated with army engineers, might organize them into a mapmaking organization like the one that was eventually to be created, the Ordnance Survey: But Smith was not interested, or the board not interested in him, and a once-exciting-sounding project that would have been ideal for Smith eventually came to naught.

But, most significantly hurtful of all, there came the formal foundation in 1807 of a new and vitally important London learned association, the Geological Society—and with it the wounding realization that despite all his evident contribution to the science, William Smith had pointedly not been invited to join. The impact of that decision was to rumble thunderously down the years—and it was to culminate, as we shall see, in the most delicious of ironies; but for the moment, in 1807, the news, when it was transmitted to Buckingham Street by friends, must have made it seem to Smith as if the whole basis of his professional existence was being lost.

✢

His personal life, too, suddenly seemed to be spiraling out of control. He was so short of funds that he began to consider selling his property to keep himself afloat. What little land he still had left from his family back in Oxfordshire had already gone. The Trim Bridge offices were now gone, rented to another tenant. All that remained that he could call his own was his mortgaged home at Tucking Mill. But try as he might, it wouldn't sell; and the owner of the mortgage, Charles Conolly of Midford Castle, was making it abundantly clear to Smith that he would not release him from the mortgage debt or buy back the house himself.

✠

It was at about this time in Smith's life that he made what appears to have been another woefully bad decision—and that was to get married. A sensible and ordered marriage might of course have been a good thing; but from all the available evidence—and there is very little; much seems to have been destroyed, and perhaps deliberately—it seems that his union was anything but sensible and ordered.

We know little about his wife, other than her name, Mary Ann. No records have so far come to light about her origins—except for the assumption, calculated from her death record, that she had been born in 1791 or 1792—or of the date of the wedding, or when (since it seems unlikely there was in fact any ceremonial) he became formally married. Everything that can be deduced stems from the fact that his diary for the year 1808 is missing, and that there are occasional references to "M.A." in the journals for many of the years that follow.

The first mention of what is probably her comes in a diary entry written in August 1809. Smith was in Norfolk, from where he wrote a note in faint pencil: "said per M.A. letter to come away Tuesday evening" and did indeed arrive in on that day—apparently telling his wife to expect him. On September 26 he records paying "Mrs. K. two months wages, £2, M.A. £2." In April 1811 he notes that he "wrote a letter to M.A." from Bath. In December 1815, more ominously, "At home all day with M.A., taken ill with pain in head."

In 1815, if her birth date is correct, Mary Ann Smith would have been twenty-four—and yet already perhaps seven years married. Smith seems to have started to live with her, almost certainly married, when she was only seventeen. And by such accounts as exist, it was a terrible marriage, with his wife neither educated nor stable enough (following the 1815 headaches, which began to worsen and transmute into more sinister ailments) to have been much of a support during these grim years.

Indeed, Mary Ann seems to have been little more than a burden to Smith for most of his life. Her death, which happened some long time after that of her husband, came about in the very saddest of circumstances. During her life she fell victim to all manner of illnesses, physical and mental. Her case notes, kept in an asylum where she was later lodged, record her as having suffered from many things—not least the pathological need for sexual intercourse that is a common-enough side effect of some of the more florid mental illnesses, and which was known then, as it still is now, as nymphomania.

Meanwhile Smith continued to try to have something—anything—published. He got into his head the notion that he might write a short description of Norfolk, a county whose geology he knew well; and he took advantage of a parliamentary election campaign in Norwich to circulate among the crowds of voters, offering them cards advertising the forthcoming book, and talking up the impending publication of his greater works. But the Norfolk booklet or map never appeared either. Smith was losing his sense of self-esteem and self-worth, suddenly reckoning himself a presumptuous intruder into the learned world—remembering that he was merely a yeoman and an orphan.

"I could previously write some sentences very well," he was to reminisce, as an excuse, "but never with sufficient confidence, and I often found a difficulty in carrying on properly the continuity of a subject."

And so those nightmare years continued, from 1806 until 1812—six miserable years when nothing seemed to go right for him, when his friends were beginning to desert him, when his muse had apparently fled, his money was trickling away at an alarming rate, and the only thing he knew how to do was to travel, take samples, make notes, and cram into his grizzled head more and ever more information about the underside of England. It fascinated and drove him still; but there seemed no future in it, for him or for his supporters, allies, and backers.

Old Dr. Richard Warner of Bath chose this time, most inappositely, to publish a guide to the city and its surroundings, making ample use of the geological information provided by his neighbor Smith, and to which he had referred generously when he wrote a Bath history ten years before. But on this occasion, in 1811, he stole freely from his onetime friend, drew a map that was an almost exact copy of one of Smith's earliest experiments, and made no acknowledgement of him at all—another example of the plagiarizing and pilfering to which the poor stratigrapher was having to become accustomed.

But then one old friend did eventually prove to be true. The Reverend Joseph Townsend, in whose house Smith had originally dictated his famous table of strata, published a book in 1812 with the curious title *The Character of Moses Established*—which set out to prove, scientifically, that the Mosaic view of the world's creation, in which of course the Flood, the Deluge, figured prominently, was still the right one. No matter the prelate's clouded scientific vision: The important feature of *Moses* is that it included, at length and in great detail, the essence of William Smith's work. It showed him to have a unique vision of the underside of the earth. It gave him all the credit that had been due and, because Townsend was a well-known and well-respected man, it placed William Smith back in the minds of the nation's chattering classes.

And whether it was because of this book or merely coincident with its publication, one key figure in this story suddenly and enthusiastically—and unexpectedly—came back into the Smiths' life.

John Cary, the country's foremost cartographer, announced that he would print and publish William Smith's great map. The two men had met some twenty years before, when Smith was working on the survey of the coal canal in Somerset. Now, without warning, this distinguished mapmaker suddenly agreed to come on board—a move that everyone knew would now guarantee that this extraordinarily ambitious work would at last see the light of day. Why Cary agreed no one knows: It is convenient to assume, however, that he came across Townsend's curious book, and was impressed by the evident importance of a man who he remembered well from their canal-making days. Yet the reason matters much less than the outcome: In the blink of an eye, Smith's fortunes were changed, and, on this singular occasion, very much for the better.

The first task was for Cary—working at his offices at 181 Strand—to make a wholly new topographic map of England and Wales. On this William Smith—who was working nearby on the floor of the main dining room at 15 Buckingham Street—would superimpose his geological information—information that he had now spent more than fifteen years, and hundreds of thousands of lonely miles accumulating. Smith decreed from the start that the scale should be five miles to the inch—meaning that the map itself, if fitted up in one sheet, would measure 105 inches by 74 inches, or 8 feet 9 inches high by 6 feet 2 inches across.

Smith had no doubts at all: What he was making was going to be mightily impressive—it was going to be a grand, grand map, big, eyecatching, memorable, and entirely appropriate to the majesty of the topic that it would picture. It was, after all, the first-ever map of the geology of an entire country, and not just a country but the most important kingdom, as Britain saw itself, on the face of the civilized planet. There was no other such map of an entire country anywhere in existence. What was being created in London was to be the model that the rest of the nations would have to follow. This map was to be a world leader, in every conceivable way—and it was to look magnificent too, so that no one would ever have cause or occasion to doubt its excellence.

Cary performed the outline drawing; one of Smith's lawyer

friends, Henry Jermyn of the ancient Cistercian abbey at Sibton, in Suffolk, was chosen to help perform the engraving, which included the writing of place-names, numbering in the thousands. It was evidently a pleasurable time. "I have a copy of Cary's map spread out on the carpet," wrote Smith, "he turned to his valuable collection of old authors—and thus did we proceed in marking the names . . . in those gleams of light thrown on the dark pages of our history we had many pleasant discussions."

The task was performed with infinite care, and yet at a rollicking pace. By February 1813, of the sixteen copper plates* on which the topographical map was being engraved, three had already been finished. A year later and the remaining thirteen were done, and Smith could now begin the equally painstaking task of colouring the individual strata, the key ingredient of a geological map. As he explained in his diary on the day he started, it was his wish "to render the map as interesting as possible to those who are desirous of knowing all they can of their country. . . ."

He now knew by heart, from all his years of travelling, the geology of a handsome proportion of England's secondary rocks. His knowledge of the Jurassic, for example, provided him with a core of stratigraphical information that allowed him to colour in lines of strata from the south coast of Dorset to the east coast of Yorkshire—he could provide a swathe of information that more than justified his ambition to create a map of the whole country. His time in Norfolk had allowed him to add details of the Cretaceous, too. To the oolites and clays of the Jurassic he could now add his new and formidable knowledge of the Chalk—a rock which, together with the honey-hued limestones of the Cotswolds, is perhaps the country's most distinctive outcrop, a potent symbol no less (in the White Cliffs of

*Fifteen of the plates were for the map itself and a sixteenth reserved for a small bonus, a geological cross-section of the country from London to Snowdon, showing the strata inclined at an angle exaggerated to underline the phenomenon.

Dover) of the insular nature of the island kingdom.

On April 18, 1814, Smith was interested indeed when he passed Cary's shop in the Strand, and saw that in the bow windows of the store were four of his sheets completely finished, and fully coloured. Cary had chosen to surprise and delight the forty-five-year-old mapmaker—he had placed the finished sheets (which included sheet XI, the area around Bath)—in the window without telling anyone. Smith, who became as excited as a schoolboy, snatched up the sheets and immediately—his first reaction, and a noble one—took them over to Somerset House to show to Sir Joseph Banks. Sir Joseph had been the project's first supporter. He had given Smith fifty pounds to begin the subscription. And yes, maybe he had harrumphed with exasperated impatience eight years before, when there seemed no end to the project in sight. But now all was coming to fruition, and, Smith reasoned, Banks should and would (especially as a near-neighbor) be the first to see the finished sheets. If he agreed, then he would be named on the completed full-scale map as the person to whom the entire project had been dedicated—the person to whom Smith owed most of all.

Banks agreed readily—for the map, he declared, was a most handsome thing. He would tell his friends, urge them to offer their support even at this late stage. And so it came—six weeks later, in June, and William Smith was being summoned by no less a figure than Lord Hardwicke to present the completed sheets to the leaders of the Board of Agriculture. They agreed wholeheartedly with Sir Joseph: the map was already magnificent, and once finished, excellent, in every way.

Smith was told to make sure that he sent a prospectus for the map to every member—and he promptly wrote as polite a solicitation as could be imagined, indicating that

> William Smith will explain the Subject of the Strata at his house, 15 Buckingham Street, the Strand, on this and the following days between the hours of eleven and five, to such

> gentlemen as choose to subscribe towards the publication of this great national work. W. Smith's Discoveries of Regularities in the Strata, with their accompanying organic remains, will be illustrated with Engravings of his large Collection of Fossils, which are placed in the same order as they lay in the Earth.

Implicit in this letter was a powerful sales pitch. Everyone who could buy one of these maps, an example of (as purchasers were not allowed to forget) this *great national work,* should now sign up to do so. It mattered little how much people could afford: there was, as John Cary had decreed, a convenient spectrum of tariffs. All depended on whether a buyer preferred the keenly priced edition, with the sixteen unmounted paper sheets and a memoir all bound up in a box of stiff blue board for five pounds; whether he opted for the standard full-size canvas-mounted map, ready for placing on a wall, and which would set him back seven pounds; or whether he decided to splash out and buy the deluxe version, the edition that was offered on canvas, mounted, varnished, set on spring rollers and issued with a fitted leather carrying case—for which the price was the not inconsiderable sum of twelve pounds. Cary and Smith were open to all kinds of financial models. And William Smith reckoned, with Cary's implicit agreement, that he should make about twenty-five shillings on every copy sold.

Ten months later and all was quite ready. On March 15, 1815, we have him writing in his journal that he had "finished corrections of five western sheets of the Map, which now completes the general corrections required. . . ."

It remained only for three crowning moments.

The first came in mid-March 1815, when the prime minister himself, the great reactionary figure of Lord Liverpool, came to Buckingham Street to inspect the immense map; he pronounced himself well pleased, and congratulated Smith. That, considering Smith's background, was no small feat indeed.

The second came later that spring, in May, when Smith went to the Society of Arts and formally presented to its president, the duke of Norfolk, the completed map for inspection. Smith, stony broke as always, was at this time only too well aware that thirteen years beforehand the Society had offered a prize of fifty guineas* for the first mineralogical map of the nation. He was eager that the Society should now inspect his work, and decide whether to pay up: they did, two weeks later. Smith records the moment matter-of-factly in his diary entry for June 16—"Received from Dr. Taylor, Secretary to the Society of Arts, their premium of £52.10s.0d. for my Mineralogical Map of England." Guineas, the money of the upper classes, clearly did nothing for him.

And the third crowning moment came on August 1, 1815—the official date for the formal publication of the map, its distribution to those who had subscribed, its offering to bookshops around the land. The coloring was all done and dried; all copies were now numbered and signed, and in most cases had been colored, by the hand of William Smith.

It was then, and it remains now, a truly magnificent thing—huge, beautiful, and filled with absorbing and elegantly managed detail. And, by comparison with modern maps of the geology of the country, in the very broadest sense, uncannily *right*. It is also, as well as being a scientific document without peer, tremendously attractive as a piece of art. It is highly colored in a way that mimics the colors of the rocks below, so that one can almost imagine that Smith was painting a portrait of the country, with all the foliage and topsoil stripped from it, such that only the rocks remain—green for the chalk, blue for the Lias, and a honey-colored, orange-hued bright streak of evening sunshine for the outcrop of the Middle Jurassic that he loved so much.

The map is entitled with suitable grandiloquence, "A Delineation of The Strata of England and Wales with part of

*The society left it up to the winner whether to take the money or a medal. For the eternally impoverished William Smith, it was no contest.

Scotland; exhibiting the Collieries and Mines; the Marshes and Fen Lands originally Overflowed by the Sea; and the Varieties of Soil according to the Variations in the Sub Strata; illustrated by the Most Descriptive Names."

And there is the dedication too, as promised, to the man who stood by him—impatiently betimes, but who had stood true until the end. "To the Right Honble Sir Joseph Banks, Bart., F.R.S., this Map is by permission most respectfully dedicated by his much obliged servant, W. Smith." And there is the date, "Augst 1, 1815." As John Phillips, his nephew and first biographer, was to write: "From that hour the fame of its author as a great original discoverer in English geology was secured."

Four hundred copies of the map were printed, numbered, and signed. About forty of them are known to remain in existence. Collectors today regard them as valuable beyond price. Huge sums are commanded for those very few copies that come onto the market. They are renowned within the rarefied world of the map dealer. There are several copies in London. One of the finest is the one that hangs behind the blue curtain beside the main staircase of the Geological Society of London, in Burlington House, Piccadilly.

✠

It is a concealment that abets a kind of shame. For considering what happened next, and considering how members of the Geological Society of the day dealt so cruelly with this man who had created so magnificent a testament to the science they were following, it is remarkable that the map is permitted in the building at all. It is in a way a haunting reminder, rarely seen now, of the way in which scientists, and especially British scientists of an era long past, behaved quite unforgivably toward one who was so self-evidently not one of their own.

For although the map that was about to change the world had now at last been published, and although the great and the good of the land—the prime minister himself included—had seen it and marveled at it and had pronounced it a wonderful creation, the man who had made it was about to begin the most wretched of all the times of his life. And it was men who even now were professing themselves to be geologists who were almost wholly to blame.

13

An Ungentlemanly Act

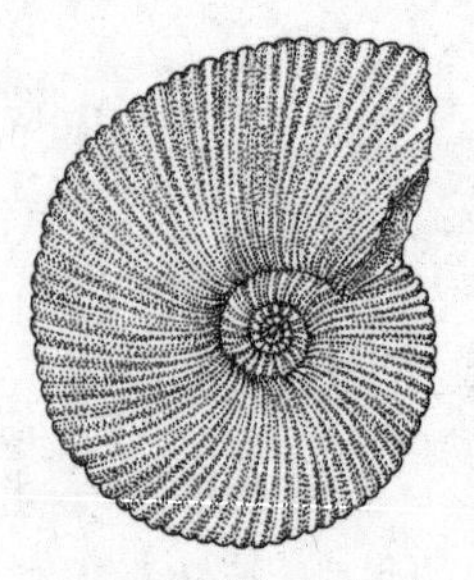

Macrocephalites macrocephalus

The roots of the tragedy that were to befall William Smith go back some years before the moment during which he briefly savored his great cartographic triumph. It was a morning during the early spring of 1808, seven years before publication day, when a small delegation of worthy and distinguished Londoners arrived at the front door of William Smith's riverside house on Buckingham Street. He had invited them there to make an official inspection tour, to see for themselves two things that were currently setting all scientific London on fire—and this at a time when geology, suddenly, was a new, exciting, and very fashionable science.

They had been asked first to see the enormous collection of fossils that Smith had brought up from his former home in Bath and had cleverly arranged "in the order of their appearance in the strata," as he put it, in cabinets in his otherwise empty dining room. And they had been invited also—or maybe this was their own idea; the records do not say—to make, at least ostensibly, a critical appraisal of the great new geological map on which Smith was then said to be engaged.

If they had secondary reasons for examining the map, they did not let Smith know. It was for him in any case an important and an intimidating encounter—a formal examination of his half-made map by members of the newly instituted Geological Society of London, of which as yet William Smith had not been invited to become a member. It was crucial, at least from Smith's point of view, that the meeting went well: Not only could the society's members help him—they could also welcome him into their congenial midst, and accelerate his progress into the very center of Britain's geological establishment.

The visiting party was led by a young man of undoubted distinction. His name was George Bellas Greenough, and the lacquered brougham that brought him and his companions to the Strand had come from the House of Commons, where he was, though only thirty, a sitting member for a comfortable Surrey constituency. He was a man of immense wealth, and was at the time building for himself a large Italianate mansion on the fashionable west side of Nash's newly laid-out Regent's Park.

Everything about Greenough, from his dandyish clothes and comportment to his formal manner and conversation, marked him as a gentleman. He was in rank and reputation as different from William Smith as it was possible to be. Yet few knew the source of his family fortune, and the man who would be the first president of the new-formed Geological Society (and was for the time being chairman, a post long since abolished) was at pains to conceal his ungentlemanly origins: In fact his maternal grandfather had been in trade as a quack apothecary, and Greenough's Liver Pills were a popular remedy for a range of maladies from chronic flatulence to simple low spirits. But they made him a fortune, all of which was in due course passed on to George, while he was a student at Eton, when the grandfather died. There had been only one condition attached to the will—that though George had been born a Bellas, his pill-maker-grandfather's gift was accompanied by a request that the boy bear the Greenough name in perpetuity.

George Bellas Greenough was an archetype of the small group of men that had come together, six months before, with the aim of creating a social and dining club to be called the Geological Society of London—"for the purpose," as the original manifesto proclaimed, "of making geologists acquainted with each other, of stimulating their zeal, of inducing them to adopt one nomenclature, of facilitating the communication of new facts, and of contributing to the advancement of geological sciences."

The society today is immense—it has nine thousand members—and is generally reckoned to be authoritative, academically rigorous, and contentedly catholic in its membership. Back when the society was formed at the Freemason's Tavern, on November 13, 1807, it was anything but. Its thirteen founding members—"we are forming a little talking geological Dinner Club, of which I hope you will be a member," went the invitation written by Sir Humphry Davy—were first and foremost cultured dilettantes (though admittedly visionary ones). Most of them were wealthy, all were possessors (though mostly for their value as modish drawing-room accessories) of cabinets of fossils and collections of pretty minerals, and in the fields of geology and mineralogy, all except one were rank—if leisured—amateurs.

Four of them were doctors. Three were chemists (one of them was William Allen, cofounder of the firm of Allen & Hanbury, which still makes cough lozenges; another was Davy, the isolator of sodium and potassium and inventor of the miners' safety lamp). Two were printers and booksellers (these were the brothers, Richard and William Phillips—the latter being the society's actual founder). One was a minister in the Unitarian Church. (At least several of the other founders were Quakers—suggesting a degree of freedom from the intellectual stranglehold of religious dogma, even in those early days.) And there was one wealthy and entirely independent man: George Bellas Greenough.

The only one who was a practicing mineralogist—geology

had generally been called mineralogy until the middle of the eighteenth century—was Jacques-Louis, comte de Bournon, a Frenchman who had fled to London during the Terror, had changed his name to James Lewis, and had established a profitable business organizing and classifying the mineral and fossil collections of London's great and good (as well as laying out the diamond collection of the fabulously wealthy collector, Sir Abraham Hume).

Of this group at least the Unitarian minister, a man named Arthur Aikin, knew William Smith and his work. Aikin was himself something of an amateur cartographer, and had a good knowledge of the topography of Shropshire and of its outcrops of minerals. There had been a halfhearted attempt by John Farey to bring the two men together, in the hope of speeding up the progress of the map—it came to nothing, however. But it was quite probably Aikin who proposed to his brother members that, having received the invitation from Smith, they go down to Buckingham Street and see exactly what he was doing—even though, quite pointedly, they had not invited him to join the society in the first place.

It seems appalling and cruel, from this viewpoint, that Smith was not seized upon as an ideal member of the early society. But a roll call of men who were offered membership in those first few years indicates why: Greenough, Aikin, Babington, Pepys, and their colleague founders wanted men very much like themselves to join their "little talking dinner club"—elegantly distinguished men such as those who were speedily admitted, men like Lords Oriel, Seymour, and Seaforth; the bishop of Carlisle; Sir James Hall, Sir Abraham Hume, Sir Thomas Sutton, and (the soon-to-be) Sir Francis Beaufort; the Right Honorable George Knox; David Ricardo (later an MP); the Reverends Edward Burrow, Matthew Raine, George Sampson; and—the greatest humiliation of all for Smith—his old friend and benefactor from Bath, the Reverend Benjamin Richardson.

But Smith himself simply would not do. In the eyes of the

Greenoughs and Sir James Halls of the world, he was unpolished and ill educated. He did not know how to dress or to dine. His accent had the common and rounded vowels of Oxfordshire. His efforts, however laudable, had not rendered him sufficiently wealthy—he would probably find it difficult to manage the fees and certainly would not wish to spend the regular fifteen shillings charged for a society dinner. Neither, it was noted with asperity, had he married well enough* to counter the unfortunate circumstances of his rustic birth.

The simple fact that he was dependent for his living on the practical applications of geology—on drainage and surveying and the holding back of the sea—meant that he was wholly unsuitable to mingle with men who liked to debate in contented languor over the competing virtues of Neptunism and Plutonism, or who looked at fossils for their beauty rather than for their usefulness in determining, as Smith did, which rocks were older than others.

Nor, it was charged, was Smith an admirer of Abraham Werner, the Saxon geologist whose theory that all rocks had been precipitated from the sea—Neptunism—had convinced many of the founders of the London-based society. There was a good chance, Greenough and his colleagues believed, that Smith had never in fact heard of Werner—a further indication of the rough-hewn, artless ways of the man, which would sit ill with the manners of such learned figures as had already been asked to join. The fact that Wernerian theory was soon shown to be arrant nonsense cut little ice: His methods—the so-called *continental* methods—were those followed by a large number of those London geologists who thought of themselves as *au courant*: Rude provincials like Smith were condemned by the Wernerian elite, and generally regarded with contempt.

*It was probably shortly after this meeting that William Smith actually married; his wife, Mary Ann, far from helping him socially, became in very many ways a considerable burden.

And if all this were not enough—why, Smith was a friend and protégé of Sir Joseph Banks! And Banks had argued long and loud that the Geological Society should not be separate from the Royal Society itself, the grandfather of all London learned societies, and the one of which Banks was still the president. He tried to change the Geological Society's constitution and, when he failed, resigned his membership and walked off in a huff. Certainly, Greenough and Hall were to say, no friend of Sir Joseph Banks should thus find it easy to become a member of their dinner club—particularly if he was socially unacceptable and not a true believer in the ways of Herr Werner of Freiberg.

There was tension, then, when Greenough jangled the doorbell and Mrs. Kitten allowed the members of the delegation inside. Smith was ready, and brought the party smartly upstairs to see his work.

It was all a terrible disappointment. The men spent only a short while looking, muttering among themselves, offering Smith merely curt pleasantries. They seemed not to be in the slightest bit impressed. In fact, they appeared almost bored.

Smith was candor itself.

"I scrupled not," he wrote in his diary,

> to explain to these gentlemen (I think rather too freely) the order of the strata and the use of fossils so arranged as vouchers of the facts, not knowing but that the new body, the Geological Society, might be inclined to serve me. In all probability the maps were also opened and explained.
>
> I was rather surprised that Sir James Hall could find nothing in such an extensive collection which seemed to please him. A time was fixed for another visit—they came, but I was not to be seen.*

*His diplomatic invisibility was, we must suppose, of his own choosing.

That first visit had resulted in what was, all told, a humiliating encounter. John Phillips, later reading between the lines of Smith's melancholy diary entry for the day, notes scathingly that the visitors offered him "only paltriness and condescension, such as they might have offered a grocer." They walked out within the hour, leaving Smith with the firm impression that he could expect neither help nor sustenance from the society, nor would he ever be invited for membership. The meeting convinced Smith that rigid class distinction lurked deep within the very science of which he was a practitioner—"the theory of geology is in the possession of one class of men, the practice in another."

That was a polite way of explaining to himself and his friend that he had been most cruelly snubbed. A terrible wrong had been done to him; it would be nearly a quarter of a century before the Geological Society, by then purged of its dilettante beginnings, realized the error and acted, dramatically, to make amends.

But there turned out to be far more darkness and menace about Hall and Greenough's behavior on that spring day than their mere contemptuous dismissal of their host. Precisely coincident with the encounter with Smith, they embarked on the plan that was eventually to destine him to ruin. They decided—and it is not unthinkable to suppose that they did so in Greenough's brougham as it clip-clopped back from the Adelphi down to Parliament that lunchtime—that they themselves should create a great new geological map. They would harness the efforts and expertise of the entire society and its growing corps of members. They would make a huge map that would become the official, definitive geological portrait of the nation, and would thereby deal a fatal blow to any further hopes of the ignorant rustic who had been so impertinent as to dare make one first.

Moreover, in constructing their new map they could, if they played their cards shrewdly, have access to the one impeccable source of the information they needed: William Smith's great map itself. They could find additional geological data elsewhere, of course; but the simplest and most direct way to discover the core material needed for a new map—though no one would ever say such a thing openly—would be to *copy* the very work that Smith was doing. They could turn themselves, unbeknown to Smith, into the very "scientific pilferers" that he had long suspected were working against him. They could become, in short, cartographic plagiarists.

And in short order, this is what—under the leadership of George Bellas Greenough—they became. They saw to it—by means unspecified but certainly devious—that copies of Smith's work fell into their hands, they pored over them, they traced the lines of the strata onto outline maps they had acquired for themselves, and then, year by year, they added information of their own, such that their end result would have the appearance and the utility of something entirely new.

✢

Politics, adversity, and misery, it is said, each make strange bedfellows. William Smith was to feel the sting of all three—and among his opponents were strange bedfellows indeed. None was stranger, perhaps, than John Farey, his old friend and pupil from Woburn, who was to become an unwitting conspirator in the devious process that unwound. His involvement began when Greenough—who was no mapmaker, and not really much of a geologist either—became stuck, unable to work out the best way to start work on the map, and turned to Farey for help.

He did so because, right from the start, his plans for the actual making of his map began to go awry. The cause? Greenough's basic notion that his map should be entirely empirical, and that its drawing should be based only on observations and not tied to any particular theory about which rocks were where, and why and when and how they had been laid down. Rumination, Greenough reasoned, had no place in a geological map: What should appear on the finished sheets of his chart should reflect facts that were quite unsullied by any theoretical presuppositions.

There should be nothing on the map, for example, that hinted at any supposed importance of fossils—nothing that suggested, as William Smith was suggesting, that certain rocks could be identified by the fossils contained in them, and that intelligent deductions could be made about the relative ages of the rocks that were so identified. Nothing of this sort was to be allowed on the Greenough map: Following the teaching of those Europeans who clung to the idea that all rocks were precipitated from marine solution, only the results of observation were to be engraved onto the map. Maybe theories would result once the finished product was out there for all to see—but for the time being, observations only.

It was a cartographic process that was doomed from the very start—for by denying any knowledge of which rock horizon might be coeval with any other, it swiftly became demonstrably impossible even to think of linking the representation on the map of any one outcrop with any other. And so Greenough, not too proud to concede that he had made a mistake, turned to the one man he knew personally, and who he suspected might help him—Smith's old pupil John Farey.

The request—which in normal circumstances Farey might have turned down out of loyalty—came at an appropriate moment. Farey was at the time working on his Derbyshire survey. William Smith, it will be recalled, had recently fallen out with him, for a collection of trifling reasons—he felt that he himself should have won the commission offered to his old pupil; he had reacted petulantly when it was Farey, not him, who was chosen to supply entries on geology for Rees's new encyclopedia; he was furious when references to his own great work on canal surveying were left out of the same book by the editors. Farey knew only too well that Smith was angry with him—and it would be fascinating to learn that this knowledge contributed in any way to his decision to help Greenough. There is no written record: One can only wonder.

But help Greenough John Farey most certainly did. He happily showed Greenough samples of his mentor's early works, and explained how it was all being done. He urged Greenough to pay little attention to the European scholars and to Abraham Werner, who made their primitive maps of Europe simply by showing rocks in their arrangement of different *types*. He would be far better off if he followed the unfolding doctrines of William Smith, who matched the strata to their unique assemblages of fossils, and was able to draw maps showing rocks according to their different relative *ages*. And to illustrate the point to this languid and leisured amateur, Farey pulled out map after map after map—all the new, unpublished, and unprotected work of the man whom Greenough had publicly and cruelly dismissed.

It was a ghastly error. Despite their recent falling-out, Smith had always trusted Farey. He had allowed him to copy his early maps, believing they would be used by him privately, never dreaming they would find their way into the planning of a rival publication. When he found out what had happened, he wrote bitterly in a letter to a friend describing what he saw, quite rightly, as yet another slight:

> Farey, it seems . . . they thought best to convert into a friend, and he either lent or gave them a one-sheet map of the stratification of the island—a copy, I think he told me, of the uncolored manuscript one that he had before given to me. And of course, his copy of Cary's large map on which he had so many years before been drawing the lines of strata was also made use of . . . . And now, as a specimen of the liberality of the leaders of public bodies (for such bodies are generally led by two or three men), I may observe that . . . they deal with Mr. Farey, by making [him an] honorary member* and

*It is one of the more curious aspects of this story that, notwithstanding all the help he gave Greenough, Farey was not awarded membership in the society either. So Smith was wrong in suggesting, enviously, that he had an honorary membership: In fact he had none at all.

neglecting me—purposely, it would seem, the better to suit their sinister views.

Armed with an abundance of advice—which, though it had come from Farey, had in essence all come from Smith—the Geological Society began its twelve years of hard grind to produce the rival sheets. Although a Committee on Maps was formally set up in 1809 to oversee the project, it was widely acknowledged within the society that it was really Greenough himself, with the help of his boundless personal fortune, who helped see the map to a successful conclusion.

Because of this the rivalry that ensued became highly personal: The battle over the map was as much a fight between William Smith and George Bellas Greenough as it was a battle between Smith and the society, or between Smith and whatever was thought of as the geological establishment of the day. It was a contest too between the ways of a man who was not afraid to get his hands dirty and the ways of a perfumed *flâneur*.

Smith was the epitome of the practical man, always grubbing around in the earth, draining fields, building watermills,* corresponding with engineers about pumping projects,† descending deep into coal mines. Greenough, on the other hand, was a young man of leisure whose works were largely confined to the library or the drawing room, and yet who was blazing a trail for what would soon be reckoned the most exciting and intellectu-

*Recently discovered correspondence shows that between 1801 and 1804 he helped to build an extraordinary mill on the River Yeo in Somerset. It had an immense underground wheel fed by water that traveled along a twelve-hundred-foot-long tunnel. Smith visited the farm in 1805 to inspect the mill's workings and, not unreasonably, to collect payment.

†Smith engaged for many years in a lively correspondence with the famous Cornish steam-engine maker Richard Trevithick, about using his pumps for the various coal mines in which Smith was working; and he was involved in the breaking of a huge underground obstruction in the Thames, where tunnelers were working at Blackwall. His range of interests was prodigious.

ally vital scientific discussion club in Europe. The great map battle can be seen in today's light as a conflict between early geology's doers and its thinkers, between the men of the hammer and the men of the quill—not, it has to be added, that George Bellas Greenough was ever known for his intellect; he was always thought to have a second-rate mind but first-rate connections.

Greenough aside, this was above all a battle that need never have taken place, and it had ruinous consequences that the Geological Society was later to regret, mightily.

Now that he had worked out how to make his map—*more along Smith's lines, and not those of the Europeans*, Greenough would have said through gritted teeth—it remained simply for him to acquire the geological data for the engravers. These he acquired from three sources. He read everything he could lay his hands on; he accumulated scores of notebooks full of information as a result of sending out hundreds of official leaflets called *Geological Inquiries*; and he went, with his colleagues, into the field.

The first two means invariably brought him back to Smith—many of the small number of English-language books and pamphlets mentioned Smith and his techniques; and scores of the replies from men who responded to the *Inquiries* leaflets also mentioning his name. One woman, our redoubtable Etheldred Bennett of Wiltshire, wrote in response, and her letter is quite typical of what must have been so galling to the bumptious society chairman: "I never yet have been able to get any information here regarding the Crockerton Clay," she wrote, ". . . but Sir Charles Blagden . . . informed me that Smith told him it was a hump in the bed of clay beneath the green sand."

The third method of collecting information was to go into the field—but Greenough's excursions were very different from Smith's field trips, which involved the mapmaker stopping coaches, jumping down into the mud, racing off to hit things with hammers and collect specimens and go down shafts and measure dips and strikes and perform all the classical tasks of a common geologist. Greenough, on the other hand, wrote in a letter to a "Mrs. S" from Richmond in Yorkshire, that he would strike out into the countryside, and

> as soon as we arrive at an inn half the inhabitants of the place are put in requisition—innkeepers, waiters, ostlers, postillions, wellsinkers, masons, gamekeepers, mole-catchers—and these are catechized one after another and, if their accounts vary, are confronted and cross-examined. Thus we soon become possessed of a vast deal of local information—which we string together as we can and then determine how much to take on trust and how much must be verified by our own personal investigation.

It was not a method much appreciated by the more pedantic and less gentlemanly of his fellow scientists. When one early sheet—of Westmoreland—was published in advance of the main map, one critic, the geologist Thomas Webster, said that "Greenough's map I found so very defective and inaccurate that I was obliged to begin *de novo*." Another critic, named Underwood, who wrote to Webster as the project reached its conclusion, was even more vituperative:

> Greenough considers England as done. *This coxcomb's reign must soon be over* [emphasis added]. The day is gone when a man could pass as a geologist in consequence of having rattled over 2 or 3000 miles in a postchaise and noted down the answers of paviors, road-menders, brick-makers and lime-burners. Slow and careful investigation and a profound knowledge of Mineralogy and Zoology must henceforth characterize the productions of the Geologist.

Failing that, however, there was of course, as crib sheet, William Smith's map itself—for the first edition of the great work came out in August 1815, by which time Greenough had still not finished his own work. There is plenty of evidence that he saw and studied the map, circumstantial evidence that he copied from it. Among Greenough's papers held in the Geological Society archives today there are no fewer than four copies of Smith's completed map; on one of them there are annotations in Greenough's distinctive handwriting—"this sheet of no further use to the Geological Map," he had written—leading critics to make the obvious inference that he took all the information that he could from Smith's map and pasted it onto his own. Without, it has to be said, any acknowledgment at all.

Now that he had the information, it remained only to make the finished map. In February 1813—by then he was probably aware that Smith really was going to finish his task—Greenough paid to have a base map engraved. This work was finished in the summer of 1815, with the entire map—at almost exactly the same scale as Smith's—on six copper sheets. By October 1817 the mountains and hills had been engraved; by January 1819, the title.

Finally, beginning in the spring of 1819, and using the intelligence gained from all those ostlers, paviors, and mole catchers—and all the data that had been, in the view of Smith and his allies, borrowed, purloined, copied, pilfered, plagiarized, and just plain stolen, or taken with no by-your-leave from William Smith—George Bellas Greenough began adding, with delicacy and elegance, the geological information. The lines of strata were drawn on, the colors were chosen, the shading was executed, a key was constructed, its position was decided. The underside of England and Wales, now officially determined and demarcated by the thirteen-year-old Geological Society of London, was about to be published.

It duly came out, published by Longman's and distributed by a bookseller named Smith on the Strand (who was of course no relation to William Smith). Greenough, aware that Smith's wall map, which was not selling well, was priced at seven pounds, decided to undercut him: He would still make money, he reasoned, if he sold his new map for six guineas, and to members of the society, whether they were ordinary, honorary, or foreign, it should be just five guineas. Undercutting Smith had an immediate and devastating effect—and it coincided almost exactly with his committal to debtors' prison. The precise nature of cause and effect can be argued about. The coincidence of events, though, was just too cruel.

In any case the reception of Greenough's map was generally lukewarm, and it sold very poorly. In the first year after its appearance Longman recorded selling only seventy-six copies. The critical reaction to it in fact rather vindicated Smith's genius. Purchasers said, scathingly, that there was nothing very new about Greenough's work—there was, despite all the waiting and the expectation, not very much more in this map that had been produced by a society and backed by private wealth, than in the predecessor that had been published five years earlier by one impoverished man, working on his own. It might well be a useful helpmeet for people traveling to England, wrote one French geologist, A. J. M. Brochant de Villiers. But with characteristically curt Parisian dismissiveness he added, and then underlined for emphasis: "*Mais il n'y a rien là pour la Science.*" There was nothing there for science.

But whatever the details, this apparently official map was now out before the public, competitively priced, selling for substantially less money than the map on which William Smith had worked for so long, and on which he had long since founded his hopes for security and recognition. Smith had for years been in the deepest financial trouble; for months the public anticipation of the appearance of Greenough's map had so limited his own sales as to effectively ruin him.

He suffered both as a consequence of the new map's pricing and because of the confusion caused by the new map's appearance. It confused, for instance, the professor of mineralogy at Cambridge, Edward Clarke—a man whose own mind was already a wonderful confusion of interests, filled with information about passions that included zinc, the making of blowtorches, ancient Greek marbles,* old coins, the chemistry of barytes, and the history of the Cossacks. He had long been on the subscription list for Smith's map, but was then told by the Geological Society that another chart was in the works. He wrote to John Cary, Smith's publisher:

> When I allowed my name to be added to the list of subscribers to the map you mention, it was under the idea that Mr. Smith was publishing the Map of the Strata of England for which the Geological Society have been for so long collecting materials. But I desire it to be distinctly understood that if the map you mention be not published by the Geological Society, and under their auspices, I do not intend to be a subscriber; and in that case Mr. Smith must excuse me for declining the purchase of his map; because I really cannot afford it.

When Smith's friends rallied to his side and accused the society's president† of theft, Greenough issued a bleating statement of apology. He had been charged, he wrote, with

> trespassing upon ground which I knew to be by right of preoccupancy, his [Smith's]. . . . The two maps agree in many respects, not because the one has been copied from the other, but because both are correct; and they differ in many, not from an unworthy apprehension of my part of being deemed a plagiarist, but because it is impossible that the views, the

*His obituary suggests that, Cambridge or not, he was not averse to cunning: In order to obtain one particularly celebrated specimen, his obituarist notes that he "bribed the *waiwode* of Athens, purchased the statue and obtained a *firman*."

†Greenough had been made the society's president in 1811, the term *chairman* then being allowed to lapse.

opportunities and the reasonings of two persons engaged on the same subject should be invariably the same.

Out of what the Geological Society tactlessly called "professional courtesy," William Smith was sent a copy of the map that Greenough had, in the view of many, stolen from him. He was presented with it at the cheap inn in which he was living in Ferrybridge in Yorkshire, where he was taking a respite from the financial trials that had so afflicted him in the capital. He was almost overwhelmed with financial and domestic difficulties—he had no money, his wife was evidently going insane, he had no job and no apparent future.

And now this.

"The copy," he later wrote in his attempted autobiography, "seemed like the ghost of my old map, intruding on business and retirement, and mocking me in the disappointments of a science with which I could scarcely be in temper. It was put out of sight."

✣

Nearly half a century was to pass before the society finally made full amends. They did so in 1865. William Smith himself was long dead. An entirely new generation of geologists now held the reins in Burlington House—men and women who were simply devoted to the betterment of the science, and who were in no sense class obsessed, were not dilettantes, were not given to drawing-room pleasantries and dandyish conceits. Their immediate predecessors had already gone some way to restoring Smith's reputation. But in the specific matter of the map the society members of 1865 recognized there was one further step they could take: On all further editions of the society map, it was unanimously agreed, were to appear the following words: "A Geological Map of England and Wales, by G. B. Greenough, Esq., FRS (on the basis of the original map of Wm. Smith, 1815)."

14

The Sale of the Century

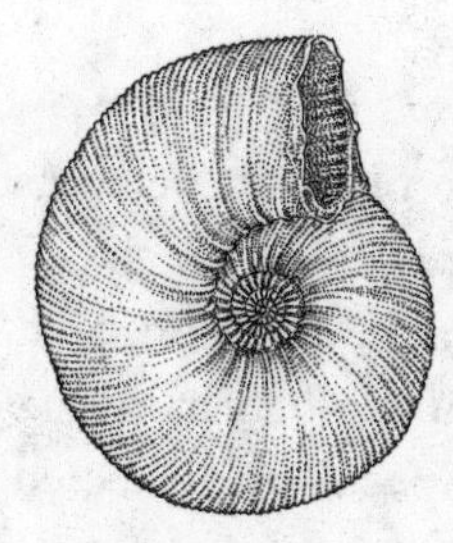

Sigaloceras calloviense

Well away from the throngs of tourists, in a highly secure back room the size of an aircraft hangar, on the third floor of a modern annex to the great Victorian palace in South Kensington that is London's Natural History Museum, are more than sixty rows of cabinets, many of them locked with double doors of steel. Inside the cabinets are hundreds upon hundreds of low and wide mahogany-faced drawers, each labeled, each with polished brass handles. From a distance the drawers look as though they should be in a safe-deposit vault and hold diamonds or the deeds of houses, or maybe an immense collection of antique maps or naval charts.

In fact, as befits the biggest natural history museum in Europe, they contain fossils—more than nine million fossils ranging from the most complex animals and plants ever known to the merest smudges of discoloration that are said to be the first single-celled hints of ancient life. In the drawers are countless specimens, perfectly selected, impeccably maintained, of once-living matter of all ages and of all types, making up what is possibly the finest and most definitive collection of paleontological remains anywhere in the world. Researchers come from all over the world to consult what is still an ever-expanding collection, doing so under the congenial invigilation of the staff of specialist curators whose job it is to manage and protect one of Britain's most important and yet unsung national treasures.

Almost all the fossils have been taken from their original collections, examined and classified and placed—most usefully for researchers—in boxes and drawers according to their age and type. All trilobites have been placed *here*, all ammonites *there*, all Jurassic dinosaurs in *that* section, all Cretaceous echinoids in *those* drawers, all gymnosperms and conifers and hexacorals in *these*.

There are a very few collections, however, that have been allowed to stay as they were and always have been—collections that have been made by truly great scientists and which are important today both for the individual fossils they contain, and, notably, for the overarching historic importance of their having been assembled together in the first place. Specimens collected and kept by Charles Darwin are to be found in the museum drawers, as are the entire amassments of early geologists like Sir Roderick Murchison, the fossil sponges of James Bowerbank, and the plants and cetacean bones of James Sowerby,* as well as some of the amazing Lyme Regis monsters eased from the Dorset Jurassic by Mary Anning.

One of these collections, which still has all the perfect integrity—the same handwritten labels, the same pen-and-ink sketches, the same accompanying notes—as it had when it was handed over, was that which now rests in the eighteenth row of the cabinets on the museum's third-floor repository, and is so large and

*James Sowerby is quite possibly the only scientist to have both a flower and a whale named after him.

extensive that it occupies no fewer than twenty-nine drawers. There are twenty-four drawers full of fossils and a further five drawers of interesting rocks—all of them carefully amassed and fastidiously cataloged during thirty years of working and traveling by William Smith.

Except that there is a difference: While the collections of scientists like Darwin and Sowerby and Murchison came usually to the museum as part of a bequest or as a gift, and at the end of their collector's life, William Smith *sold* his fossils to the museum, and did so while he was still a middle-aged man.

No matter that he had so painstakingly and single-handedly collected all 2,657 of them. It counted for little that his study of them and where they had come from formed the foundation of his theories, of his writings, of his great maps and of the entire science of stratigraphy that he had helped invent. It seemed to be of no consequence that these thousands of fossils were every bit as important to Smith as though they were part of his very person. The fact is that he sold them all, and he let them vanish into the vaults of the British Museum and he did so because, quite simply, he had to. For by the time he handed them over, in 1818, William Smith was a deeply desperate man.

Ever since the turn of the century he had been in financial trouble. A freelance life—which is what he had chosen for himself after being dismissed by the Somerset Coal Canal—inevitably dictates a precarious existence. In Smith's case, it need not have been so. Smith was very good at his job: whether he was called in to make a survey, to drain a field, to prospect for coal, to shore up coastal defenses, to repair hot springs, or build a mill, or install a winding engine, he performed his task superbly, easily earning the two or three guineas that he charged each day, each time adding to his reputation, each time making fresh contacts with men who would give him more and yet more work. Moreover, he had some rentable property: the long-ago death of his father meant that he had title to his old cottage and the neighboring field in Churchill, and when his uncle William had died in 1805, he came into some nearby lands: from both he managed to win an income of about £100 a year. Added to his professional payments his income was at times, very considerable, and might well have kept him in funds indefinitely.

But Smith was a spendthrift, too—spending in large measure in an attempt to gild the impression he would make on the noblemen for whom he so often worked. Not that he spent money on fine clothing, or on entertainment, or wine. His particular vice seems to have been the collection of good addresses—the purchase of the estate at Tucking Mill House, the office at Trim Bridge, the town house on Buckingham Street, all grander and more costly premises than necessary, but all with the sufficiency of style and élan that meant he could ask the likes of Sir Joseph Banks to come calling, could entertain the Chairman of the Geological Society, or have the Prime Minister himself drop round for tea.

The costs of the ceaseless traveling needed for making the map mounted over the years—although in the early days some of these expenses were borne by his clients, as he piggy-backed some of his map work on commissions he won, quite reasonably. "Everyone who travels," he wrote, "knows that ready money must be provided for the road. There is no credit at the coach offices. A man's hand is constantly draining his pocket, and so pressing for fees were all the lackey attendants. . . that I used to say a civil answer could scarcely be obtained on the road for less than sixpence. In taxes and tolls alone the man obliged to travel much pays heavily, however abstemious and economical he may be. No man may take less on the road than I did: for, 20 years since, I used to go from London to Bath without tasting anything."*

*In his diary for 1804 Smith gives the post-chaise fares when he and an assistant journeyed through Yorkshire—Hull to Weighton (of the Jurassic's notable *Market Weighton Axis*, q.v): £1.5s.11d; Weighton to York: 11s.9d; York to Tadcaster, 9s.4d; Tadcaster to Spofforth, 16s.9d; Spofforth to Leeds, £1. 8s.6d—a total of £4.12s.3d outlaid, before a penny piece was spent on hotel rooms, guides, or food.

To compound his woes, and in the saddest of ironies, he also made one celebrated and unpardonable geological error, which was to cost him dear.

In 1807, hoping to supplement his income with a steady flow of funds without having to do much work, he bought himself an old quarry on Combe Down, one of the hills above his home at Tucking Mill. The idea was to prise out of this quarry chunks of high-quality Bath stone—the oolitic limestone then so favored for building across the country (and which he knew was just then being used to refurbish Henry VIII's tomb in Westminster Abbey). He would take these immense pieces on a small specially built light railway down to the coal canal, have them band-sawn into building stones, and ship them off on barges to the Kennet & Avon Canal and thence via the Thames to the masons' shops in London.

It took him some while to get the scheme off the ground. To buy the new land he needed to sell a small part of the Tucking Mill estate—already he had been obliged to raise a second mortgage from the owner, a Mr. Conolly of Midford Castle, to raise the ready money needed to pay for his travels—and that sale took rather more than four years. But by 1811 he was ready. He had made £1,330 from selling part of his land, found £500 by mortgaging a little more, plowed some into paying back rent and taxes, and then paid to have the light railway built, and the stone wharves erected on the coal canal's banks. By 1814 stone was being cut, the sawmill was in full swing, and narrow-beamed barges, heavy with William Smith's oolites, were being horse-heaved slowly toward the capital.

Good fortune lasted for no more than a year. Two events—one outwardly irrelevant, the other directly related—conspired to put a sudden end to the project. Come the following June—when Smith's map was all but complete—the duke of Wellington defeated Napoleon at Waterloo, the wars at last were over, and England was forced to count the cost of all the fighting. It was a vast sum. The nation went into a prompt and deep financial tailspin. People stopped spending large sums of money: They halted building projects halfway through, they abandoned plans, they deferred decisions.

Already, in the spending of comparatively small sums, a new mood of postwar fiscal prudence had settled on the land—one can infer that much from the sparse list of paying subscribers who actually put down the relatively small sums of money that Smith was asking for copies of his map. But in the spending of larger sums—also cut back, hugely—the problem for Smith was an even greater one: For, with their new fiscal prudence, Britons who had lately been so careless about putting up grand houses and the newfangled structures known as office buildings all over the place, all of a sudden stopped doing so—and not only did they put a stop to erecting new real estate, but they also put a stop to buying stone with which to build it. And that suddenly sounded the death knell for the quarry Smith owned in the hills above Combe Down.

Not that they now would have bought Smith's quarry stone anyway. From 1815 onward it had become abundantly clear that the Combe Down quarry possessed an unusual quantity of very poor stone—an outcrop of Oolite that was, in more ways than one, very decidedly Inferior. The quarry went immediately and spectacularly bust. The workers were laid off, the railway closed and left to the rust and the willow herb. And William Smith was swiftly to realize that the hundreds he had spent—and, more important, the new mortgages he had incurred—were all for naught. He was spiraling out of control, and it seemed that nothing could be done for him.

And not only was he now married, to a woman who was by all accounts going spectacularly crazy, he had also taken in tow a relative—a young man who would be his apprentice, helpmeet, and student for most of the rest of his days. He has been mentioned in passing before: He was called John Phillips, and he was

William Smith's orphaned teenage nephew.* The fact that he would one day go on to be professor of geology of Oxford University, and president of the Geological Society—astonishing and delightfully ironic testimony to Smith's magic as a teacher and kindly mentor—meant little as the storms began to gather: In 1815 John Phillips was, to put it bluntly, a financial burden to Smith too. His school fees—thirty pounds a year—and his food and clothing bills can only have added to his uncle's savage decline in fortunes.

Twice during 1814 the bailiffs had been to the Buckingham Street doors, hammering for settlement, demanding back taxes and unpaid rent, threatening prison. Twice during that year Sir Joseph Banks, now convinced that the map he had helped sponsor truly was going to be published, had bailed Smith out with a check and a word in the ear of the vexed collectors. Now, come 1815, with the map almost done, with Combe Down collapsed, with John Phillips in residence, with Conolly of Midford Castle pressing for funds, the situation was far worse than even Sir Joseph could imagine. Smith knew there was only one course of action left open to him, and with this Banks himself concurred: He must sell his fossils, and if not privately, then to His Majesty's Government.

The negotiations for the sale, which Smith initiated the summer of 1815, turned out to be painful, humiliating, and protracted in the extreme. They began when Smith asked for help from a Welsh farmer, James Brogden, for whom he had done

---

*John Phillips—who later went on to write the only full-length biography of his uncle—was born on Christmas Day 1800 to Smith's sister, Elizabeth, and her husband, John Phillips, an excise man. The boy was just seven when his father died; when his mother died six months later the boy was taken in by his uncle John. By 1813, though, William Smith, despite his straitened circumstances, was already paying John's fees at boarding school, and by 1815 he agreed to look after him full-time. The boy was fascinated with geology—Benjamin Richardson of Bath had helped instill the fascination—and within days Smith's diary records his nephew hard at work: "23 November, with JP arranging shells according to Linneus. . . . 6 February Making out Cornbrash and Forest Marble Fossils with JP. . . ."

some drainage work near Kidwelly. Brogden was now a member of Parliament and the chairman of the Commons Ways and Means Committee; Smith told him the situation and asked if the government might be able to help.

A steady stream of officials then began to call at the Adelphi house to see what was on offer. The fact that the prime minister had visited the house that same Easter, and that Sir Joseph Banks was telling everyone in earshot of Smith's eminence, must have helped, for the visitors were a distinguished group: Nicholas Vansittart, the chancellor of the Exchequer, came by, as did a wealthy landowner named Sir John Stanley, and one of the secretaries of the Treasury, Charles Arbuthnot. (Smith went back soon after to see Vansittart at his office in Downing Street—his diary shows that he gazed longingly at the huge belemnites in the flagstones lying outside the waiting-room window.)

A few days later William Huskisson,* who had the grand title "Surveyor-General of Woods and Forests," made his inspection; as did another MP friend of Smith's (who had employed him on restoring a mill in Kent) with the improbable name of Mr. Barne Barne. The conclusion of all these visitors was that Smith should write a formal memorial† to the very Gilbertian-sounding lords commissioners of the Treasury, pleading his case.

This he did in mid-July—hastily cutting and pasting nine already printed paragraphs from the brochure that was to accompany his map, and adding simply an exhortation that the government "purchase my Collection for the British Museum‡ which would add very much in point of utility to the Collection now

---

*Huskisson had the mournful distinction in 1830 of being the first man killed by a railway locomotive. He was elderly and, being accident prone, was lame from a dislocated ankle when he attended the formal opening of the Liverpool & Manchester Railway: He was unable to get out of the way when the engine, the *Dart*, bore down on him.

†Memorandum.

‡As it was then; the Natural History Museum, one of its divisions, would later be built some miles west of the main museum headquarters in Bloomsbury.

deposited there and would moreover I trust afford me means of giving to the Public a Mass of information which would be found highly interesting and beneficial to almost every Class of Society."

There was at first no reply. He waited and waited. By late September, Smith's creditors were becoming less and less patient. He wrote again, imploring the Treasury to help by advancing him five or seven hundred pounds (a not unreasonable sum, he thought, since in 1810 Parliament had voted to pay thirteen *thousand* pounds to a well-connected aristocrat, Charles Greville, for his collection of allegedly rare and valuable minerals). Still nothing happened; except that in October yet another group of officials and experts (including a chemist called Charles Hatchett who had discovered a metallic element and named it columbium) descended on Buckingham Street, where they found Smith and his fourteen-year-old nephew vigorously cleaning the specimens and readying them for valuation.

Then at last, in mid-October, the government agreed in principle to buy the specimens, and while waiting for a formal valuation, offered Smith an "imprest"—an advance—of the niggardly sum of a hundred pounds. Smith bit his tongue, remained respectful, and became firm friends (through the good offices once again of Sir Joseph Banks, whose beneficence during Smith's long and tragic story cannot be gainsaid) with an even more influential Treasury official named William Lowndes—who had by chance become a fellow of the Geological Society two years before. His friendship turned out to be the key: Over the coming months, as a torrent of letters and memoranda swirled around inside the Treasury—to and from and about a tireless William Smith, working even on Christmas Day, cleaning and cataloging the collection to make the precious fossils more acceptable to the government—Lowndes's influence, and its limits, became clear: Smith, it was finally decided, was to receive the sum of five hundred pounds, payable in four installments.

It was a bitter pill, and Smith was deeply disappointed. He had worked so very hard to persuade the government that his fossils were worth more—three times as much as they had offered, at the very least. He expected also to be paid for a 118-page catalog he and young John Phillips were creating as a handbook to the collection. But the Treasury was not interested in offering more money, and Smith had to spend yet more funds to have this slim volume published privately.

So he and John Phillips then packed up the fossils and, probably with very great sadness, stacked them ready for collection. It took several more months for curators to ready a room—but they never did collect the fossils. Eventually in June 1816, a year after opening the first correspondence with Brogden, it was Smith himself, making three round-trip journeys from the Strand to Bloomsbury, who finally carried the specimens—the 2,657 fossils, of a total of 693 different species—to their new home.

His plight briefly attracted the attention of the House of Commons. The member for Oxford, a Mr. Lockhart, mourned the "extreme parsimony" with which the government appeared to treat men of science, particularly "a gentleman who by his own discoveries had laid the foundations of geological science, who had for 30 years been cultivating that study, and had proved its application to many purposes of public utility."

But in the same fulminating speech, Lockhart reminds us all of the limited view that was held by even the most enthusiastic supporters of William Smith and his findings. "The truth of Mosaic history," he said, "is incontestably proved by Mr. Smith's geological observations . . . [which provided] proof of the Account of the Creation and Deluge in the Sacred Book."

In fact they did no such thing; though no doubt neither Smith nor his supporters would cavil over a point that, in any event, certainly reflected the prevailing mood of the day. The Bible's teachings were even then still being accepted literally—regarded as Gospel truth—by the huge majority of both the public and their legislators: The more rational views of the learned few had still not found widespread acceptance. The more important lesson to be learned from Lockhart's speech is in any case altogether another one: William Smith had, for what it was worth, an ally in the House, and his desperate situation was at least being talked about in the most important forum in the land.

In the short term such publicity did Smith and his fossils precious little good. And magnificent though the collection is still universally acknowledged to be, Smith's fossils have been seen all too little in the almost two centuries since. At first the museum authorities let them languish in a damp storeroom, with few visitors. They balked at the extra expense of putting the specimens in the specially built glass cases they had promised, but instead had Smith come round and arrange them on his old sets of sloping shelves, which he had been asked to carry in from Buckingham Street—a service for which, adroitly, he charged an extra hundred pounds. Then everything was moved, more prominently, to the South Sea Room, then back to another room so neglected that when Smith visited late in life he found them hidden behind piles of junk.

(Matters took a distinct turn for the better much later, in 1880, when the museum itself was moved to its present site in South Kensington. By that time Smith's reputation had soared, and his collection was placed prominently in a gallery of its own, with dozens of his own hand-colored illustrations in the huge glass display cases. Smith's marble bust dominated the exhibition, and for eighty years his heroic status was commemorated in style. But then again his reputation began to dim, his name became less readily remembered, the details of his curious life unknown. The fossils went back into storage in the late 1970s and remain there now, in a series of drawers, consulted by specialists but otherwise, like Smith himself, banished to a situation of respectable but little-deserved obscurity.)

The funds that were brought in by the sale and his subsequent work on the catalog proved only the shortest-lived of palliatives. Funds went out as swiftly as they came in. After one payment of one hundred pounds, Smith promptly sent twenty-five pounds to his brother back in Wiltshire, and then made out a check for a further fifty-five pounds to settle a legal action against him, most probably for debt. And yet none of what was done—no sale, no payment, no loan, no visit to the pawnbroker—could do more than briefly divert the inevitable, could do more than delay the day of reckoning.

By the end of 1816 William Smith had no fossil collection—and yet still, more or less, he had no money either. The situation had been alleviated only temporarily: A deep crisis, which was to deepen progressively over the next three years, was soon utterly to engulf him.

✢

Out of these sad times, however, came two books that added mightily to the impressive list of publications that William Smith was to leave for posterity. Both relate to the specific use of fossils: The first, a four-part work published in 1816 under the title *Strata Identified by Organized Fossils*, contains nineteen beautiful hand-colored engravings of fossils. William Lowndes, proving himself the firmest of friends, contributed fifty pounds toward its publication. The second, confusingly called *The Stratigraphical System of Organized Fossils*, and published a year later, is essentially a catalog of the "3,000 specimens . . . in the British Museum," as the title page put it, with all the hyperbole to which publishers are prey.

Neither did any good. Neither book was to help keep at bay the small army of bailiffs, beadles, process servers, apparitors, tipstaffs, and summoners who were now circling, vulturelike, around William Smith's embattled person. He had suffered much over the preceding years. The clouds had long been building. Now came the deluge.

15

The Wrath of Leviathan

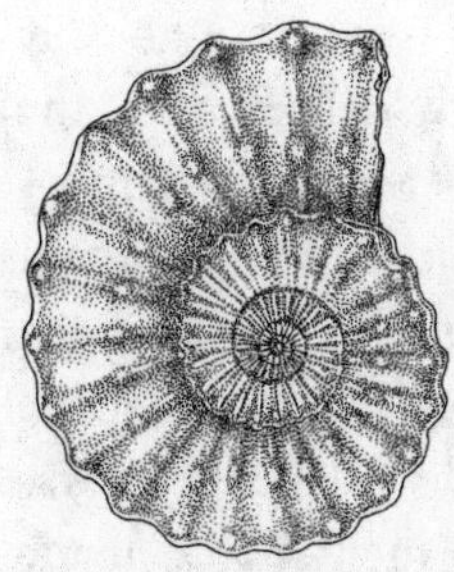

Peltoceras athleta

The storm was slow to break, but fierce and unremitting when it came. The creditors had been closing in on him for years. The dunning men were always hammering at the door, for bills unsettled, for mortgages overdue, for taxes owed but never paid. Now, by the beginning of 1818, thirty months after the publication of the great map, and at a time when William Smith should by rights have been showered with dignities and rewards, came the first grim mention of the ultimate penalty, debtors' prison. James Brogden, the kindly MP who had tried to help Smith sell his fossil collection, was the first to make public his own anxiety, in an urgent letter written in the middle of January.

"Poor Smith's distress and importunity force me again to intrude this unpleasant subject on your notice," he wrote to the Treasury officials. "The annexed paper contains a statement of what he thinks are his claims . . . if you can settle his claim today and make a small advance on it, you will save him from arrest and enable him to resume his useful employment in the country."

The Treasury got the message. An internal memo sped from one official to another: "Smith the Map Maker (I mean the Strata & Fossil Man) hopes to receive from the bounty of the Treasury some aid; I heard last night in the House of Commons that if he doesn't get that aid he will be arrested tomorrow."*

The urgency of the appeal got through: A check for thirty pounds was sent round to Buckingham Street the following morning. The bailiff on standby went away. The turnkey was stood down. William Smith was not compelled to appear before the bench, on that cold January day at least.

He continued to work and to travel, notwithstanding the financial fuss. When next the Treasury wrote, on February 12, he was away in Monmouthshire. The letter, from an official named George Harrison, had some moderately good news, though only in a tiding-over sense: Providing that Smith delivered to the museum the very last part of his collection, a cache of specimens he had been hoarding for such a rainy day as this, a further payment of a hundred pounds would be granted. He had been handed thirty pounds in January; and so when he arrived at the Treasury on March 28, he collected only the balance that was due—seventy pounds. This was the last payment to be made from His Majesty's Government. There was no more in the kitty. So far as the current and pressing problem of debt was concerned, William Smith's luck was now fast running out.

He had had the chance to cut and run, twice, no less. Anyone other than Smith might well have taken one or other of the chances—both of them exceptional, well suited to his talent. That he in the end refused both offers opened him up, yet again, to still more perfidy and ill treatment.

In the first instance he was invited to apply for a job in Russia, via no less an agency than the great Count Alexei Orlov, one of the heroes of the Russian campaign against Napoleon, and an

*The words were triple-underscored in the original.

adviser to the emperor, Alexander I. Count Orlov had come to England on a diplomatic mission and to take the waters at Bath; while there he instructed an aide, a certain Dr. Hamel, to make an official request for a British mineralogist to come to the southern part of the Russian Empire—most probably the Ukraine—to supervise the immense coal industry there.

One of Smith's old friends from Bath, the Reverend Richard Warner,* told Hamel about Smith, and for some weeks in 1815 the two men wrote to each other, with the Russian clearly offering Smith a career of wealth and honor. However, Smith, at the time arranging his fossils for the British Museum, mentioned the possibility of going to Russia to his contacts there—prompting them to respond that, provided that he turned the foreign offer down, he could be assured of an offer of a permanent and lucrative career with the museum.

Not unnaturally he immediately told Dr. Hamel and, via him, the visiting Russian count, that he regretted he could not take the job. The museum, far from reacting with pleasure and a consequent generosity of spirit, remained stiff and silent. No further offer was ever forthcoming. Thus was the disappointment that plagued Smith's life reinforced yet again.

A year later still he was offered another job, this time in the United States. He was asked, according to a note in his diary (though by whom we do not know) "to survey, level, estimate and report on the practicability of overcoming the falls on navigable rivers in North Carolina." But this offer he declined as well, without explanation. Knowing what history allows us to know today, and what would shortly befall him, it beggars belief that this troubled, unappreciated man did not slip out quietly on a westbound ship, heading for an infinitely better life in the New World. As it is, and whether by commission or default his writ-

---

*Warner was a friend despite his having copied Smith's first circular geological map of Bath for a publication of his own.

ings do not indicate, he opted to stay put in England and stare down the gathering crisis.

The law in relation to debt in London in the early years of the nineteenth century was clear enough—though by today's lights, somewhat peculiar. Very basically put, a creditor who failed to get satisfaction from his debtor could apply to a court official—a bailiff, a man equipped with extraordinary summary powers—to have the debtor committed to one of the great London debtors' prisons—the Fleet, the Marshalsea, or the King's Bench. The former was on the east side of what is now Farringdon Street, near Saint Paul's Cathedral; the two latter were across the Thames in the prison-ghetto of Southwark.*

The peculiar aspect of the law as seen today is that there seems to be no logic or sense in imprisoning a man for debt—for then he is most certainly not going to be able to earn money to pay the debt off. But the custom of English common law had been savagely firm for centuries: A debtor taken *in execution*—which was the rather overly dramatic legal phrase then used for debt arrest—was to be kept *in salva et stricta custodia* until the debt was paid.

The explanation for such a necessity is simple enough, and stems from the basic principle of the system, which held that by arresting a debtor, his body had been seized in lieu of his estates. Until he had so arranged these estates that his creditor could be paid and satisfied, then his body would be entrusted to the safekeeping of a prison warden—hence *in salva et stricta custodia*—as guarantee. There was no concept of guilt or innocence in a debt prison: Rather, imprisonment for debt was seen as evidence of *failure*—the failure of a debtor to pay what was owed, the failure of a creditor to obtain what he was owed.

---

*In Southwark there were as many as a dozen prisons; and although Dickens made the Marshalsea and the King's Bench infamous, the one Southwark jail that truly infiltrated the English language, and gave it an eponym, was the one destroyed in the Gordon Riots of 1780—the Clink.

Legal scholars argue that the subsidiary intention of the ancient laws on indebtedness was simply to preserve a system of credit. Imprisonment was threatened as a deterrent, a means of ensuring that the credit necessary for commerce—the signing of bills to raise cash for custom, for example—continued to flow easily. Men of means would sign bills, it was argued, if they knew that those for whom they signed them would be imprisoned if the bill went unpaid.

But the system was willfully misused, by both creditors and their debtors—and it took a long while for more civilized laws to take effect. Only in 1759 did Parliament rule that a trader who was entirely without assets might declare himself bankrupt and go free; and in that case he was allowed ten pounds' worth of personal goods—bedding, clothes, tools—to ensure that he still had both dignity and the means of livelihood left. If he declared bankruptcy fraudulently, however, he would be sentenced to hang.

At the time William Smith became mired in indebtedness, any more general reform of the debtor-prison system was still several years away. And when John Berisford, a noted advocate of the reformist agitations in the seventeenth century, wrote of "this great and terrible Leviathan, who thus disturbs our peace and cracks the sinews of the body politic," he could have been writing just as easily about the mountain of debt that ensnared the nation, as of the lamentable state of some the prisons in which the debtors were placed.

The magnificent copperplate script that appears at the bottom of the page of the Commitment Book of the King's Bench Prison for the year 1819—below the records of the imprisonment of the equally unfortunate Thomas Mackrell, Charles Dibdin (the Younger), Mary Ann Russell, and Thomas Smith is brief indeed. All it says is: "No. 19, William Smith, Ent&^{d} 11th June 1819 in discharge of his bail at the suit of Charles Conolly Esq., Oath £300 and upw^{d}, and was thereupon comm^{d}. By C. Abbott."

If there is to be a villain in this piece, then it was Charles Conolly. He was the owner of Midford Castle in Somerset, and the holder of Smith's mortgage at Tucking Mill House. Conolly had been financially involved with Smith since 1798, when he helped him to buy the house and small estate at Tucking Mill. He would have been involved in the refinancing that Smith had to arrange when buying the Combe Down quarry, and he had born the initial burden when, in 1815, Smith's misjudgment (or bad luck) as to the quality of the oolite precipitated the venture's collapse.

He must have tried countless times during the following years to win some of the money back: But Smith's financial recklessness would have put paid to any hopes of his doing so. Inevitably he is cast as the Scrooge of the story, as the pitiless landlord who consigned one who should have been a hero to years of humiliation and ruin. But then again, one can hardly blame a man who had been owed a formidable sum of money—the "£300 and upw[d]" of the suit—for twenty years, and now saw no prospect of it ever being repaid, heading off in exasperation to plead before the King's Bench for satisfaction—even if that satisfaction meant no more than to have Smith committed to prison and his goods sold by fiat of the court.

Smith records all too dryly, with the detachment one might find in a geological fieldbook, the events leading up to the day of his arrest.

On May 15 he was looking over another nearby house in Buckingham Street, number 12, which was up for rent—it might have been cheaper; he might have wondered if he should move or knew that he might soon have to.

On June 2 he sent a report—perhaps about the parlous state of his finances—to a firm called Dosse & Co; his friend Brogden had come around a little earlier to tell him he "should move something for me this day in the House of Commons."

On the fourth he was at the Court of Common Pleas in

Westminster Hall, face to face with Conolly at last, debating the merits of the latter's various claims. The next day Smith "made out accounts and sent them to Sir G. Alderson." I found no G. Alderson listed, but there was a Sir *John* Alderson, reporter to the King's Bench at precisely this time—and it may well have been that this judge demanded of Smith that he produce his accounts to show how severe his financial situation was.

One thing was clear, however. Smith was in no sense a trader—a dealer in goods and chattels. Had he been so he might have claimed protection under the 1759 bankruptcy legislation, as all traders owing more than one hundred pounds had a right to do. As it was, the judgment of the Court of Common Pleas was clear enough: Smith was found to be an insolvent debtor, and Conolly could have him committed to prison forthwith, and see that enough of his goods were sold so as to make up the sum that the court agreed was owed.

The next day finds Smith working on coloring more of the county maps he had in the making. No mention is made in his diary of the court, of the debt, of any sense of impending doom. But two days later, early on the cool and clear morning of Friday* June 11, 1819, the ax for which all had been grimly waiting finally fell.

The bailiff duly arrived outside the door of the Buckingham Street house, armed with Conolly's court-approved warrant, signed and sealed by the officials at the Court of Common Pleas. There would and could have been no further argument: Smith knew he was finally compelled, by all the majesty of English law, to do as he was told.

The bailiff and his charge would have ridden in the court post-chaise, along beside the Thames (a confusing journey, dodging among the wharves, since there would not be a Thames

*Had it been a Sunday, he could not, by law, have been taken—a situation which gave rise to the phrase "a Sunday man," meaning someone who could only ever go out and about on a Sunday, for fear of arrest at any other time.

Embankment built for another forty years), and across the river either by way of what was officially called the William Pitt Bridge—Londoners called it Blackfriars, as they do today—or John Rennie's* newly completed Southwark Bridge, to the Borough. Fifteen minutes of patient trotting, and the carriage with its myrmidon and his charge would have drawn up outside the high brick walls and the formidable miseries of the King's Bench Prison.

✢

London's three great debtors' prisons were private institutions, owned by corporations, run for profit, and with men in charge as wardens who could make an income of thousands a year. For one of the many puzzlements of the entire debt-imprisonment system is that it cost money to be in prison. It cost the prisoner money to enter a debtors' prison and then to stay there—nearly two pounds to get in, a scattering of other charges (such as the six shillings and eight pence due to the tipstaff for his service at court), and payment of at least one shilling a week for an unfurnished room with the most basic kind of bedding.

If the incoming prisoner was utterly without funds, or unwilling to part with whatever he had, he would be put into a chummage—a dormitory—with fellow prisoners, with whom one lived closely together, as "chums." It usually took only a few days of such discomfort and indignity for the prisoner to be persuaded of the good sense of moving to a private room—whereupon the warden would regard him as a source of revenue, treat him well enough, and offer him the kind of privileges that made many debtors think of the King's Bench and the Fleet prisons as comfortable, hotel-like retreats, where they could seek sanctuary from the pressing problems beyond the walls. Not that they were

*This was the same John Rennie who had hired Smith in 1793, at the beginning of the survey of the Somerset Coal Canal.

The King's Bench Prison in Southwark, in which William Smith—and Dickens's Mr. Micawber—were locked up for debt.

always pinioned behind bars and brick: there was an area around all debtors' prisons called the Rules, in which such prisoners as were trusted not to abscond were allowed to live and carry on their businesses—but to be free from harassment by those to whom they owed money.

The precise circumstances of William Smith's stretch in prison can never be known. The King's Bench records are unhelpful—it was a private institution, its officers had no incentive to keep records of much more the prisoners' dates of admission and release. Smith himself has expunged from his diary all material relating to his time inside. His nephew, John Phillips, who tried to remain in the Buckingham Street house while his uncle was away—he was told by the court officials to leave, however, as the landlord promptly cancelled Smith's tenancy and bailiffs came in to seize his somewhat pathetic collection of personal property—made no mention of any predicament in the biography he was later to write. The horrors of his imprisonment and the miseries of his marriage remain the two great nonsubjects in William

Smith's recorded life: All reference to them has been expurgated from his works, all muted, hidden, or efficiently bowdlerized,* both from Smith's writings, and from all other contemporary accounts.

His prison walls were fifteen feet high, topped with rows of sharp iron spikes. It was impossible for passersby to see inside, and prisoners, even in the topmost rooms, could not see beyond the wall. Provided he paid, each man was given a cell, nine feet square. He could either remain there for his term, or socialize: And if he chose to mingle, he would likely find his fellow prisoners a congenial lot—the prison population after all comprised men and women who, because they had been loaned money, were by and large prosperous by nature and habit. Debtors' prisons were generally populated by the middle classes—not by the criminal classes or the undeserving poor. The twenty thousand prisoners for debt were, it was said, from "the better strata" of society, a phrase William Smith would have found doubly ironic.

There was ample food and plenty of grog. Visitors were allowed. Men with access to funds could have young women sent in, for pleasure and amusement. And in the courtyards of the prison, all manner of games were played—the King's Bench is known as one of the places where, thanks to the happy combination of high walls and a large population of bored and competitive men wondering what to do with them, the game of racquets was invented. Visitors might think they had entered a gymnasium or a pleasure garden: Since there was no enforced routine or attempt at punishment, inmates were free to do more or less as they pleased—and callers report on the games of ninepins going on, or wrestling matches, or on groups of inmates sitting in the sun under a tent, drinking beer.

It was in the King's Bench that Mr. Micawber reminded

*By coincidence Thomas Bowdler's edition of Shakespeare, expurgated for the sake of public decency, appeared in 1818. The eponym thus derived was first used in 1836.

David Copperfield of the debtors' maxim, about "annual income twenty pounds, annual expenditure twenty pounds ought and six, result misery." But in reality, the life that Micawber lived—with the agreeable and Gypsylike dinners in the prison cells, the evident camaraderie of two hundred men and women there cursed with similar situations, and the free-and-easy manner of the warden and his staff—was none too bad. And it has to be supposed that William Smith's eleven weeks in the prison were none too bad either—except for the shocking fact that so clever, decent, talented, and hardworking a man was incarcerated there in the first place.

Whether his by now very sickly wife visited we cannot know. Nothing in the diaries left by John Phillips records a visit by either his aunt or by Phillips himself. Nor do either Sir Joseph Banks or James Brogden ever mention traveling across the river from their offices in Somerset House and Westminster, respectively, to see their friend in Southwark. Similarly John Cary, who was on the brink of publishing a second volume of William Smith's *Geological Atlas,* seems either to have stayed away or merely remains silent on the matter of ever going down to see his mapmaker. But whether they did or did not, the ordeal was over more rapidly than Smith must have supposed when he went in.

The Commitment Book records his leaving as matter-of-factly as if he were a guest at a hotel: "Dis. 31st Augst 1819. PPer Atty." A lawyer had apparently called to say that the chief creditor, Charles Conolly, was now satisfied. Sufficient of Smith's goods and chattels had now been sold to meet the demands of the man from Midford Castle. There was no longer any need to keep Smith in custody as guarantor that this should happen. After eighty nights in one of the world's most notorious houses of incarceration, he was free to go. He was escorted from his cell to the main door, given back his belongings, and the gate swung open. He stepped out into the street, and into the throng of the curious bystanders, soon after first light.

He arrived home in Buckingham Street to find the house locked and bolted and a court official at the door. His goods, this bailiff informed him coldly, were still being organized, filed, and valued. After some while he was permitted inside to get such of his papers as the court officials deemed valueless, and to take them away. Some of his more valuable papers, he learned, had indeed been sold—but to an anonymous friend (one imagines either Joseph Banks, John Farey, or [as we shall soon see] William Fitton) who had then arranged for Smith to have them back.

There was now no further point in his staying around. He sought out his wife—who had in part as a consequence of these events fallen savagely ill and, according to contemporaries, was near-deranged. He found his nephew John Phillips, who had left a message to say where he was staying. The three of them went to Holborn, to the inn where the postilions gathered, and caught the Northern Mail, the overnight stage bound for Yorkshire.

✠

It was a trying time. His first reaction was indeed one of embitterment, and he wrote savagely of his relief at turning his back at last on a capital city in which, he insisted, he had never been made welcome, in which unhappiness had attended almost all his days. "London quitted with disgust," we have already seen him note, and with sardonic pleasure. "The cheering fields regained."

A while later and his mood had become calmer, his manner more philosophical. He wrote a long, rambling, and at times barely coherent note to himself that he titled *Difficult Times Briefly Investigated, by an Accurate Observer of Passing Events.* It reflected, with some poignancy, on the brutality of life, on the trying nature of hard times:

> Time will show that my geological labours are not properly the work of an individual in my humble circumstances, but

> such as might be generously encouraged as public works—and my journeys of so many thousands of miles a year for 20 years at great expense, had they been upon discoveries in the interior of Africa instead of England, might have been a national object.

It is his later musings on the sad situation that are more memorable. Writing some twenty years after the events, he was able, quite succinctly, to compare his own fate with that of his fossils, which he knew had never been properly displayed, and which he felt were every bit as imprisoned in the bowels of the British Museum as he had been in his nine-foot cell in the King's Bench Prison.

Little would anyone suspect, he wrote of the museum,

> that in such premises there was a prison in which these innocent tell-tales of the true history of our planet were to be immured for a term of more than 21 years. . . . The man and his fossils might be imprisoned, but his discoveries could not . . . . The collection was lost, books and papers scattered, and he was deprived of everything but the stores of his mind.

He was cheered by an old couplet he had learned in childhood: "When house and land is gone and spent/Then Learning is most excellent."

It was armed with his learning alone, a few papers, his hammer and acid bottle, magnifying glass and compass, a theodolite and a chain, and in the company of his half-crazy wife, Mary Ann, and his nineteen-year-old nephew John Phillips, that he then boarded the night coach for the Great North Road. He owned no property. He had no home. His only achievements—his map, and the atlas that would be published on the very day he traveled—brought him no income to speak of. He was penniless, homeless, out of work, and—in comparison with the gentlemen of the Geological Society—out of fashion and out of favor.

He would get down from the coach three days later in the windswept Yorkshire town of Northallerton, in an effort to rebuild his life as a nomad, to wander northern England in search of a means of earning a living. In the event, he was to spend the rest of his life in the North Country; he was to find work there and, in that uncertain labor, satisfaction of a kind.

And in due course he was also to find vindication. When, after a dozen years of the hand-to-mouth existence of a fugitive journeyman, he returned to London for a lengthy stay, he reentered his capital in triumph, and he was showered with the honors, gratitude, and recognition that had for too long been steadfastly denied. Though he did not know it when he stepped down from the Northern Mail on that bleak early autumn morning, the tide that had challenged his life was, in due course, though some long time off, assuredly going to turn.

16

The Lost and Found Man

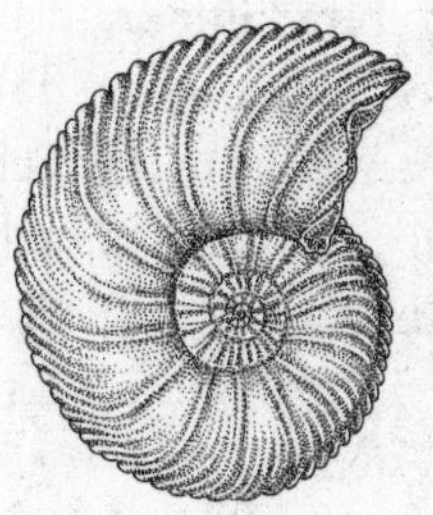

Cardioceras cordatum

In the fractious and rebellious days of late-eighteenth-century Ireland it would have been foolhardy, to say the very least, for a young man to carry in his bag a sharp hammer, a magnetic compass, and a bottle of hydrochloric acid. For William Fitton, who was an eighteen-year-old student in Dublin at the time, it was positively dangerous: The English soldiers who found him, hammer and compass in hand, and promptly arrested him on the quite reasonable suspicion of being what would later be called a Fenian subversive, took a great deal of persuading that he was merely a student of the newfangled science of geology. His explanation—that the menacing-looking items in his bag were no more than the tools of his trade—was not something the troops of Empire were eager to believe in 1798.

But then the authorities at Trinity College interceded on his behalf, and Fitton was released. He resumed his studies, behaved more prudently as the rebellions raged away outside his college walls, and by the end of the century had taken a degree. He worked only for a short while as a field geologist, in Ireland itself as well as in Wales and Cornwall. It seemed an arduous way of making a living; and in time he made his way to Edinburgh, took a degree there in the less controversial field of medicine, and after a spell in London established himself in Northampton as a doctor, specializing in the study and treatment of pneumonia.

His fascination with geology never left him, though. He joined the Geological Society, demonstrating in his papers and debates—and with his furious clashes with men like George Greenough, whom he regarded as enemies, both intellectual and social—that already the time of the dilettante was over, that science was in the ascendant, and that good men were in the process of taking charge of a new field of study that Fitton and his like believed was coming to be of truly profound importance.

It was his determination to see that good men won due acclaim that led Fitton to become the instrument through which William Smith, whom he first met in November 1817, eventually won the recognition and reputation he had long deserved. Fitton, in short, was the means by which the tides of Smith's fortune began to turn—except that by the cruelest stroke of Fate his efforts began to bear fruit only at the very moment that the poor man was about to suffer his greatest humiliation, in the King's Bench debtors' prison.

By 1815 Fitton was already a well-respected member of England's scientific establishment: He had a solid medical practice, he was a close friend of men like the near-ubiquitous Sir Joseph Banks, he had been elected to the Royal Society, and he had married well and had a considerable fortune. He was, despite being only thirty-five, a man of influence; and when in 1817 he asked to be put in touch* with William Smith, to see the nature of his work, a ripple of interest stirred the waters of intellectual London.

---

*The go-between was John Farey.

Smith's diary, which is erratic at the best of times, notes only that a meeting was arranged for early December but hints that Fitton himself never came. Others of Smith's allies did, however; and for a while afterward a small barrage of papers made their way around the learned journals of London, trying to establish Smith's credentials as the creator of stratigraphy and the creator of the map that had been on sale now for the past two years.

None of these efforts seem to have done Smith much good. His financial situation did not improve. His standing with the Geological Society did not change. His acceptance into the salons of London's intellectual *demimonde* remained on hold. John Farey had made generous mention of Smith's working progress ten years before, in the pages of the *Monthly Magazine* and the *Philosophical Magazine* and in Abraham Rees's *Cyclopaedia*—but none of these efforts seemed to have enhanced Smith's standing, and it looked more than likely that this new blizzard of panegyrics was going to be similarly ineffectual.

It was not until in the early summer of the following year that a wider public began to sit up and take notice. Only a very small number of appreciative people began to take Smith seriously—and not enough still to avert his impending disaster. But the mood, in the middle of 1818, most definitely began to change. And it did so purely because of William Fitton, because he did write his assessment—and a glowing assessment it was—of Smith's work, and he saw to it that it was published in one of the country's most respected journals of the day, the *Edinburgh Review*.

His paper, which he entitled *Notes on the History of English Geology*, was only seven pages long. But short though it was, it was a concentrated tincture of approbation. It looked almost entirely at Smith's contributions to the new science—it examined in a disinterested way, but in detail, everything of importance that had Smith's name on it—the great map, the memoir that accompanied it, the first volume of the illustrated catalog of British fossils (the second volume of which was published the day after he was released from prison), the small geological cross-section between Snowdon and London drawn on the side of the large wall map, and the first proofs of the first of Smith's seventeen large-scale county maps, which were to be published by John Cary over the coming six years, and which Fitton, cunningly and without Smith's knowledge, had managed to get to see.

The paper was an almost undiluted paean, a document that was destined (though probably not designed) to delight Smith in what was turning out to be the most troubled period of his life. Fitton had examined everything in the closest detail and written a considered criticism—noting errors, questioning judgments, criticizing lapses. But overall the tone was that of a man overjoyed by the serendipity of the experience, the enthusiastic pleasure of a man who has discovered a rare and special talent.

William Smith, Fitton told his readers, was "a most ingenious man," whose only fault was that he had been "singularly deficient in the art of introducing himself to public notice." There could be no doubt but that William Smith was performing work that constituted a truly historical development in the evolution of an entirely new science. He did not go so far as to suggest that Smith's map would change the world, but the implication in the paper was undeniably that, with the appearance of this immense and beautiful document, the world—of commerce, of industry, of agriculture, and of intellectual endeavor—would never be quite the same again.

A year later, though, Smith was a destroyed man, fleeing London for good and forever (or so he supposed), disillusioned, bitter, the victim of those he claimed were cheats and scientific pilferers. So what if William Fitton had been so kind as to write with such generosity of spirit? His doing so had made no essential difference. His fine words had manifestly buttered no parsnips. The map, now three years before the public, still hadn't sold as well as he had hoped. The quarry had gone bankrupt. The landlord had foreclosed on his mortgage and had called in a long-owed debt. The bailiffs had done their worst. He had been turfed out of the Buckingham Street house. His goods and chattels had been confiscated and sold. And now he was in Yorkshire, two hundred muddy miles away from London, and heartily glad—if, that is, he knew the meaning of gladness—to be there.

Smith was never to own a home again, never to settle anywhere other than as a tenant. As John Phillips was to write about his uncle, with the timbre of a true Victorian melodramatist:

> [F]rom this time for seven years he became a wanderer in the North of England, rarely visiting London except when drawn thither by the professional engagements which still, even in his loneliest retirement, were pressed upon him, and yielded him an irregular, contracted and fluctuating income.

It was seven years of nights at inns and coaching stations and weeks in cheap lodging houses, in towns and villages as far flung as Doncaster and Kirkby Lonsdale, Sheffield and Hesket Newmarket, Bennetthorpe and Durham Town. There was a curiously surreal, Gypsylike contentment to some of these years, which his nephew hinted at when he wrote, soon after they began their self-imposed exile:

> In the winter of 1819 Mr. Smith, having perhaps more than usual leisure, undertook to walk from Lincolnshire into Oxfordshire. According to an established custom on all such tours, he was employed in sketching parts of the road and noticing on maps the geological feature of the country.
>
> The object proposed was to pass along a particular line through the counties of Rutland, Northampton, Bedford and Oxford, but the ultimate destination was Swindon in Wiltshire.
>
> We crossed in a day's easy walk the little county of Rutland

> . . . reached the obscure village of Gretton, on the edge of Rockingham Forest . . . whatever may now be the accommodations at this village, they were very wretched in 1819, December, but the odd stories of supernatural beings and incredible frights which were narrated by the villagers assembled at the little inn greatly amused Mr. Smith.

✠

The pursuit of geology, apparently for its own sake alone, was never far from Smith's mind. "The road up Boziate Hill," his nephew continued, now writing of the pair's time in Northamptonshire,

> was mantled with fossiliferous stone, some of which obtained from the hilltop was believed to be Kelloway's Rock, and was found to contain *Ammonites sublaevis* and other fossils. A fine specimen of this ammonite was here laid by a particular tree on the roads side, as it was large and inconvenient for the pocket, according to the custom often observed by Mr. Smith, whose memory for localities was so exact that he has often, after many years, gone direct to some hoard of nature to recover his fossils.

Day after day there was the same goading restlessness about Smith—picking up, as he would for seven long years, a little work here, leaving his calling card at some great hall or grange or castle there, endlessly looking up old acquaintances in the search for work or contacts or rooms for the night, his journals always filled with references to local geology, with narratives of local lore, and lovingly noting his occasional encounters with famous men (as with Adam Sedgwick, the godfather of the Ordovician and coidentifier of the Devonian, whom he met in Kirkby Lonsdale; and the great Oxford geologist William

Buckland, whom he met on this very walking trip, and whose eccentric decency came to figure prominently, along with Sedgwick's in the next, and more gloriously culminating, chapter of Smith's life).

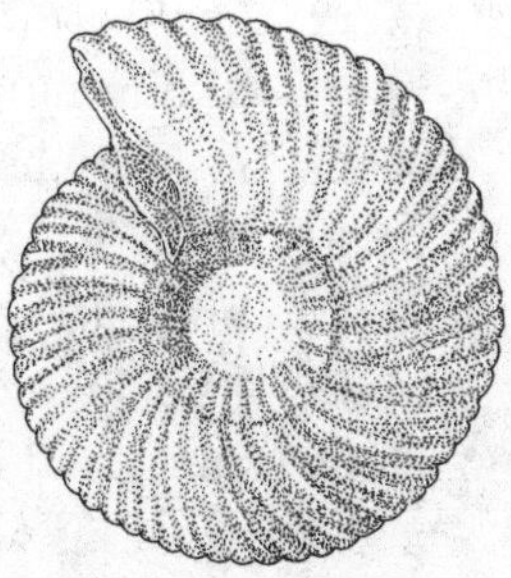

Ammonites sublaevis, from Smith's own collection, found during his tour of Northamptonshire.

There was during this period an absence of any sustaining central core to his life—there was no home, no family, no single glorious project, except for completion (which never happened) of the series of county maps. He did a little work on these each time he went south to London, where he would see John Cary, take cheap rooms in Charing Cross, and knuckle down to coloring the new information he had gleaned, onto a series of charts and cross-sections that—when seen today—appear more beautiful than anything he did before.

The cross-sections have found particular favor in the United States, where they can be seen in a popular poster published in Oklahoma. Certain of his county maps, in reproduction, can be bought today in London: But the half dozen or so that are for sale hint at another of the melancholy realities of Smith's life—that though he had the time, and presumably the energy, and though he had the assistance of a young nephew of ever-increasing ability, he did little during this period of his life of any real merit, and rarely finished the greater projects that he had started in the years before. It was almost as though he wanted to shrug off the work he had already accomplished—to live off the reputation he had won from it, but not to return to it, in case he suffered the same measure of disappointments from it that had dogged him all his life so far.

He became fond, however, of two particular places during his long sojourn in the north. One, the seaside town of Scarborough, has today a charming, if rather vague, remembrance of his long association with it. He first went there in 1820, just a year after his exile began. He had been asked by the town corporation to see if he could advise on improving the water supply—a task he accomplished with ease. But what impressed him most about the town was its setting, its fresh sea air, its obvious charm—all of which might help to improve the health of his wife, now deteriorating fast. Phillips, who rarely wrote about Mary Ann Smith, includes a line that speaks volumes for this fraught period in his uncle's life: He had come "to this romantic and delightful town in hope to soothe the mental aberration of his wife, which became very manifest in this year."

"I went to Scarborough," Smith was to write in his attempt at autobiography, "under distressing circumstances—unknown to anyone in the place but through the medical men." But then he uncovered a happy chance about the population of the town—and noted in his journal that "everyone here is very fond of talking on Geology." Within only a few weeks his temper improved; his apprehension of finding himself friendless and alone had abated. Within weeks he had won himself a circle of admiring and intelligent friends; he wrote that he found the people—provincial and unlettered Yorkshiremen though they might be, and by London standards innocent and artless folk—to be a constant delight. They had *no side* about them, he said; they accepted him for what he was, they made him welcome, they provided him with intellectual stimulus, and, most important, their evident pleasure at having so clever a man in their midst enabled him to renew a sense of self-esteem that had been soundly battered during his unhappy creative years in the capital.

He completed his detailed geological map of Yorkshire while there—it was published in 1821. He also began offering lectures—a guinea for a series of nine,* a single lecture for three

---

*"Ladies" were offered the same series for only fifteen shillings, a discount designed to lure them from their sewing rooms.

shillings—and with his nephew turned himself for a while into a highly profitable traveling road show. He pitched his tent, as it were, in York, Leeds, Hull, and Scarborough, time and again, offering the latest information and the most up-to-date arrangements of organic fossils, for the improvement of the minds of all.

It was when he was off to give a course of lectures in Sheffield—for sixty much-needed pounds—that he was first troubled by the only illness he ever seems to have had, a curse of his peculiar trade. The notes for his possible autobiography record the moment:

> On sallying out from my winter quarters on a sunny day in March, and in hammering a long time for the fossils in blocks of the Cornbrash rock at the back of the Castle Hill I so caught the rheumatizm, which was the worst complaint I ever experienced, being six weeks confined to my bed in great pain and the loss of the use of both my legs, though all the time in perfect health and good spirits.

He was determined to make his assignment in Sheffield. He had the stagecoach brought to the front door of his inn, and demanded that two men "tumble me in like a sack of potatoes." He had to support himself on his arms during the fifty miles of the journey, but sang songs to keep himself cheerful. On arriving at the Sheffield junction he tried to get up, "fell like a child," and had to be carried off in a sedan chair. He gave his lectures; they were well attended; he collected his money. But that evening he mused for the first and only time that he might perhaps now give up geology for good—rheumatic joints being the principal curse of a field explorer—and keep a school.

It was an idea that vanished as fast as it arrived. Geology still managed to exert its powerful magnetism on the man. And by now Smith was finding Scarborough so congenial that he was coming back time and again, whether for a commission or not.

He returned first for a two-year stay in 1824, lodging with a family called Williamson, who later wrote that "Smith and his eccentric wife established themselves in our house, where they dwelt for a considerable time."

✢

Before long Smith also made a very considerable physical impact on the town. He helped, most notably, to set up the Scarborough City Museum—a curious rotunda of a building that was finished in 1829, and that still stands on the town's seafront, though in much reduced condition. Smith's idea was radical: He had called for the building to be designed—it was built in the Doric style, fifty feet high and nearly forty feet in diameter, with a spiral staircase and a graceful dome—so as to allow thousands of fossils to be arranged on shelves around the outer walls, all in their proper relative positions. The younger remains, in other words, were to be displayed on shelves at the top of the building, while the older fossils were deeper down,

The rotunda-shaped Scarborough City Museum, designed by Smith for the specific purpose of allowing fossils to be seen in their proper chronological order.

nearer the base of the building. The youngest members from the Cretaceous would be at the top, the oldest fossils from the Trias would be at the bottom, all of them arranged just like the strata lying deep in the earth outside, just as the strata were arranged elsewhere in North Yorkshire, in exactly the same order as the rocks in faraway Dorset, and no doubt just as the rocks were arrayed, Cretaceous up above, Trias down below, in all the distant corners of the world beyond.

It was a beguilingly clever idea. The members of the Scarborough Philosophical Society—a number of such local institutions were being set up all around the country at the time, designed for the dissemination of knowledge about science—were all hugely enthusiastic. They quickly agreed to fund the venture, and within weeks they had backed Smith with a handsome subscription. Building started swiftly, and it was finished inside a year. There was a gala opening, with sixty for dinner, "the table being spread with every delicacy of the season, with a fine dessert and excellent wines."

The idea at first worked fabulously well, and for a while anyone who was interested in the paleontology of Yorkshire—which has some of the best fossil locales in the country, particularly rich in ammonites and dinosaurs—was compelled to make a pilgrimage to the curious little drum-shaped building by the sea, in which was housed "one of the most perfect fossil collections in England." But in recent years the building has become a tawdry and half-forgotten little structure, the only relic of Smith being an indifferent reproduction portrait, and a diorama painted around the upper floors, probably by John Phillips. There is a memorial stone on an outside wall mentioning Smith as having helped conceive the idea. There are no fossils in the Scarborough Rotunda anymore; Smith's shelves are filled merely with indifferent items of junky memorabilia.

He went off surveying and draining in the Midlands once the museum was up and running, and spent six years away. He came back again to Scarborough in 1834—and though he only rented a small house called Newborough Cottage on Bar Street, he was happy enough to put down roots, so far as he could, and to stay in the town for what turned out to be the rest of his days. A letter he wrote to his niece, Ann, survives:

> I am now busy in partly furnishing a neat cottage situated in the midst of pleasure ground and walks laid out by Marshall, the tasteful designer and author of Rural Economy. We have two parlours and a kitchen, cellar and other conveniences—three good bedrooms with two staircases and attics. Shall have possession on Monday. Rent £15 a year. You may therefore direct to me in future at Newborough Cottage, Scarboro'. The place was occupied by the late Mrs. Eastwood, aunt to Mr. Hall, who kindly undertakes to convey this intelligence to you. I shall have plenty of room to spread out MS., maps and fossils, and in this snug retreat for doctors and philosophers I shall be happy to see you and the Professor whenever you choose to come.

The letter was written in October 1835. Smith was by then a man of sixty-six—and, as the letter suggests, was slipping gracefully and—at long last, contentedly—into his old age. And *contentedly* is the key word here—for something significant by now had happened to change his mood and temper. The train of events that brought him to a new state of untroubled contentment began at the one other Yorkshire site in which he spent time, and where he experienced the epiphany that was to redirect, and for so much the better, the course of his remaining life.

The pretty village remains more or less unchanged since Smith was there during the six years from 1828 until 1834. It is no more than six miles inland from Scarborough, hidden deep in a fold of the hills, in the valley of the River Derwent. It is called Hackness, and it and the rolling hills around it have for years been the fiefdom of a family of lowland Scots called Johnstone—

Hackness Hall, the Yorkshire seat of the Johnstone family, where Smith's achievements and genius were finally recognized.

a family that would come to have an inestimable impact on the life of the old geologist who came, for a while, to stay among them.

Sir John Vanden Bempde Johnstone—there is Dutch blood liberally mixed in with the Scots—was described by his biographer as "a sincere friend to geology and natural history." He had joined the Geological Society, was a keen collector of fossils, as well as being an MP—first for Weymouth in Dorset, and then for Scarborough (spanning, in his legislative responsibilities, both ends of England's main Mesozoic outcrop). When, as a baronet—the Johnstone family of Hackness was not to become ennobled for a further eighty years*—Sir John succeeded to the

*A baronetcy was the most junior hereditary entitlement—the appellation "Sir" being handed down, according to the principles of male primogeniture, indefinitely. But it did not mean that the title holder had been ennobled: One had to become a baron to be called "Lord" and to sit in the upper house of Parliament; the ranks of viscount, earl, marquess, and duke then rose in stellar fashion, one above another. Above dukes were only princes of the blood royal; below baronets, mere knights and then the rest, known sniffily as "gentlemen."

Yorkshire estates, he decided to "convert to practical effect on his farms some of the geological and botanical truths which he knew to have been established in the museum and the laboratory."

He knew of William Smith from meeting him at the Philosophical Society meetings in Scarborough (he had contributed, *gratis*, the stone for building the rotunda), and in 1828 hired him as his land steward. He gave him the use of a vicarage close to where Hackness Grange now stands (as a hotel); and for the six following years Smith lived as pleasant a life as can be imagined, meditating, writing, living (as Phillips noted, with some asperity) like "a happy farmer," and performing only one task of geological significance—making a beautiful large-scale and fully hand-colored map of the Hackness Estate, which hangs today in pride of place in the Hall—the big house—having survived a fire that raged through most of the rest of the house early in the last century.

All passing geologists of note now suddenly took great care to stop at Hackness, both to see Smith and to congratulate his patron—this "sincere friend to geology"—for taking care of the kindly old man who bumbled amiably around his estates. One of those who visited was the Derbyshire-born chemist and lens maker William Vernon, who came by in 1826. It was precisely at this time that the first suggestions were made that William Smith's contribution to geology should somehow be formally recognized.

Whether Vernon first put the idea in Johnstone's mind, or whether (as the romantic view maintains) the enlightened nobleman actually suggested it to Vernon remains unclear. But what is known indicates the subsequent chain of events. William Vernon dashed off a letter to Roderick Murchison, the Scottish soldier who had established himself as one of the great architects of geology, the founder of the Silurian period and, with Sedgwick, cofounder of the Devonian. In this letter Vernon said, as forcefully as courtesy allowed:

> Smith has dedicated his life to geological enquiries, and has done perhaps more than any individual for the science, and is at an advanced age in poverty and dependence.
>
> There has been nothing in his conduct or character to diminish the respect due to his exertions in the cause of knowledge and the compassion which his circumstances excite . . . . I have thought a subscription might be raised . . . a small annuity purchased for him, sufficient to secure his not dying in the Poor House.
>
> I should be much obliged to you if you would do what you can to forward it. I am sure you will find many able and willing friends to this project, in Dr. Buckland and many other members of the geological Society.

The mills of the Geological Society grind exceeding slow. There was still some opposition among the old guard—the "anti-Smith alliance" of Greenough and his friends, who had seen to it that as late as 1822 Smith was still denied even membership of the organization. But this was now changing, and rapidly. A new breed of scientists was directing the society's affairs these days—scientists who accorded as much honor to the practical men, the men who went out into the field in the damp and chill and happily dirtied their hands in the finding of facts, as to the theorists and thinkers in what was, after all, a fundamentally practical field of study.

Vernon's appeal—or was it perhaps Sir John Johnstone's appeal?—fell on ears that were fully attentive. All agreed that some distinction should be given to Smith: The only question was the timing—the feeling that, given the continuing power and influence of Greenough's faction, it would be imprudent to make an immediate move. Smith was content, employed, more or less free from financial woes: Matters could wait awhile longer.

But by the time Adam Sedgwick was settled in the chair of the society four years later, it was time to move. All of a sudden, and at long, long last, the most powerful and influential body of geologists in the world could right the wrongs it had done to Smith during the past quarter century. It would summon the old man down from Yorkshire to London, it would garland him with the greatest honors it had in its power to bestow. Now was the time for the scientific establishment to beg forgiveness, and to offer to William Smith his long-overdue reward for what was officially recognized as one of the most memorable of human achievements.

17

All Honor to the Doctor

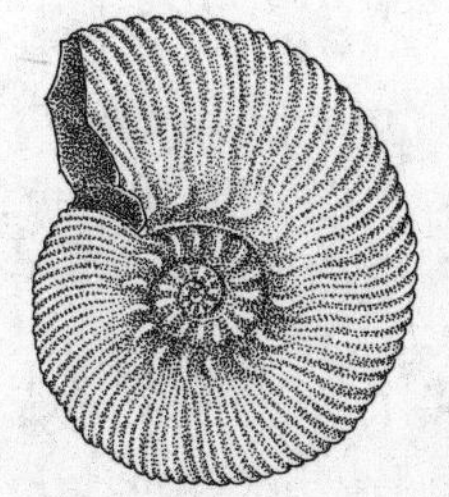

Aulacostephanoides mutabilis

"A medal shall be struck in fine gold, not exceeding the value of ten guineas, and it shall have on its obverse side the bust to the left and to the right, Wollaston."

It was just before Christmas 1828 that William Hyde Wollaston died in London at the age of sixty-two. Two weeks earlier he had written a formal letter to the Geological Society, announcing that he was planning to make a bequest of a thousand pounds' worth of stock, in the hope that an award might be created—the first year's income would be for making the die—and handed out each year to one individual for his or her research "into the mineral structure of the earth."

It was in some ways an odd choice. William Wollaston was not a geologist but a chemist, though said by all to be a man blessed with the most acute powers of observation. He could apparently see even the tiniest of flowers while riding on horseback. He invented the camera lucida* after noticing something

*A projection device that helps artists by throwing a reflected image of an object directly onto a sheet of paper, from which it can be drawn.

odd in the crack in his shaving mirror. He was one of the few men who ever noticed a mirage on the river Thames. He was a doctor, an expert on kidney stones and on mineral-based enlargements of the prostate.

He specialized in the chemistry of exotic and precious heavy metals. He discovered palladium and, from its rose-colored salts, refined rhodium. He made a fortune from finding clever new ways of working with platinum.* He invented the reflecting goniometer, an instrument that could measure the precise angles between the faces of microscopic-size crystals. And during his lifetime he published no fewer than fifty-six academic papers on "pathology, physiology, chemistry, optics, mineralogy, crystallography, astronomy, electricity, mechanics, and botany," but made not a single contribution to the science of geology itself. He has, however, a mineral named after him—a calcium silicate called *wollastonite*, which is formed when limestone is crushed hard against another rock.

But his medal, awarded every year since 1831, is without peer in the world of geology: To be a recipient of the Wollaston Medal is to become the equal of a Nobel laureate, in a discipline for which (despite its universal and elemental nature, and in common with mathematics) Alfred Nobel puzzlingly and shamefully left no bequest. The Wollaston is the Oscar of the world of rocks, fought for gamely, campaigned for bitterly, and if awarded, then accepted with the secure knowledge that career and reputation are guaranteed for life.

Yet at the time the medal was struck the science it sought to reward was still very much in its infancy. Physics, chemistry, and medicine had long been fully fledged: When Nobel offered his prizes for excellence in them, they had been around for cen-

*For much of its history the Medal has been struck from palladium, the silvery metal that Wollaston discovered and named after a passing asteroid. But as the Society's demands have occasionally exhausted supplies, gold—as in Smith's case—has had to be used instead.

turies. Wollaston's bequest, on the other hand, was designed for practitioners of a science that had been totally unknown—at least by the name *geology*—a mere six decades before. And it was a science that, because of bigotry, intolerance, churchly disapproval, and the fundamental assumptions it sought to challenge, suffered more than the usual share of growing pains—pains that were mirrored precisely by the difficult evolution of the Geological Society itself.

Set up when the science was just staggering to its feet, the society was made up initially of men who had more pretense about them than solid achievement. In the outside world, said George Greenough, "the term geology . . . was little understood." Few men "beyond the pale of our little coterie" aspired, he said, to become geologists. The early members of the society—or at least, the more powerful among them—were bent more on following field of elegant drawing-room fashion than pursuing and disseminating much rigorous scientific research. Most of them seemed blissfully unaware that there was in fact a very real and a profoundly important science buried within the aimless charms of their fossil and mineral collecting.

So there was a good deal of tension, unanticipated at the time, and yet with hindsight quite understandable, once

The Wollaston Medal, the first of which, awarded in 1831, was given to William Smith, "the father of English geology."

the men of real worth—those who recognized that they were in at the beginning of a very real new science—began to rise through the society's ranks. There was tension in particular once such men began not simply to rise but to overwhelm and displace the smug "little coterie" whose fossil collectors' dining club it originally was.

In few sciences can there have been such an ugly rivalry. John Phillips, William Smith's nephew, first came to live full-time in London in 1831 and reported that he found it all vastly amusing, noting in his journal that "the jealousy among the men of science here is wonderful." Others found it less so. Thomas Webster, who was the curator of the society's museum, wrote that he was surrounded by "a band of busy, jealous, active and revengeful witlings who have gained and kept their ascendancy partly from contempt, partly from the indolence of others." George Greenough, who as president of the society was responsible for setting its early temper, had once been held in awe: Now he was being dismissed as merely "a charlatan and a blockhead"; others in his various factions were described as "quacks," "jackals," and "pilferers."

Gradually, though, the passions subsided; and by the late 1820s the new elite was in place, to remain unassailably there for many years. Controversies continued to rage, as in any stimulating environment—but thenceforth they tended to be about geological dogma and discovery, much less about money, influence, and class. The academic tone was rising by the day: Sedgwick and Murchison, the two colossi of the discipline, were now firmly in position, battling royally over such technical matters as the succession in the Lower Paleozoic. In place too was perhaps the most colorful and clever of all nineteenth-century geologists, the great showman-scientist who preceded John Phillips as professor at Oxford, Dean William Buckland.

Buckland is perhaps better known today for his antics than his discoveries and beliefs. He had a voracious need to check everything. He tried to eat his way through the entire animal kingdom, offering mice in batter and steaks of bison and crocodile to guests at breakfast, but reserving the viler things for himself—he declared that he found mole perfectly horrible, and the only thing worse was that fat English housefly known colloquially as a bluebottle. His sense of taste seems not to have been ruined by such experimentation—he once found his carriage stranded in the nighttime fog somewhere west of London, scooped some earth from the road and tasted it and declared to his companions, with relief, "Gentlemen—Uxbridge!"

He was a great skeptic, particularly where Catholics were concerned. Once, led to a dark stain on the flagstones of an Italian cathedral, which the local prelate insisted was the newly liquefied blood of a well-known martyr, he dropped to his knees, licked the darkened spot, and announced that it was in fact the urine of bats.

But above all he was a fieldworker—a scrabbler-about in meadows, a clamberer, a hammerer. His best-known discovery was of the thousands of broken bones in a cave in North Yorkshire; these he declared to indeed be evidence of the Deluge, the Flood—but since the bones it had swept up into the cave in cold, windy Yorkshire were those of tropical animals—hyenas, elephants, lions—the event had in fact happened many thousands of years ago, when the weather was tropical, and not according to Ussher's dogmatic timetable.

The six days of Creation were properly to be thought of as Six Ages, he said. Scientific observation had now displayed, without a shadow of doubt, that there was much more to the Bible than could be learned from its strict literal interpretation. The new science of geology, in other words, was now capable of asking—and probably of answering—the truly great questions, the fathomless wonderment about God, the cosmos and humankind. The conclusions to which Buckland, the keenest of all observers, eventually came were to change the relationship between science and religion for all time. And geology, this brand-new science, with a brand-new society and now a brand-new medal from Wollaston, was the key to the unlocking of thousands of years of fettered and blinkered prejudice.

It was this holy trinity of geology's new young grandees—Sedgwick, Murchison, and Buckland, none of them older than forty-five—who saw to it that it was William Smith who would be awarded the first-ever of the society's Wollaston Medals.

No matter that he was not a member of the society and had never been invited to be, or that just nine years beforehand his membership had been blackballed by Greenough's old guard. No matter that he was lowborn, that he was uneducated, a provincial, a convicted debtor, and that even more contemptible of figures, a practical man. No matter that he had accused some of the more august members of the geological establishment of stealing his work, of cheating him of his due. No matter that a quarter of a century's worth of ill feeling had accumulated, that factions existed and battled among themselves, and that the kingdom's entire geological universe had been upset. Now, the wise men decreed, was the time for atonement and reconciliation.

The message had gone out from Sir John Johnstone at Hackness, via Mr. Vernon, to Sir Roderick Impey Murchison, from Murchison to Dean William Buckland, and from Buckland to Adam Sedgwick, who was both Woodwardian Professor of Geology at Cambridge and, since 1829, the Geological Society's president. The debate was brief, the die was cast. A formal resolution was passed at a special meeting of the society's council, without dissent (despite Greenough's council membership) on January 11, 1831, both that the first medal be given to Smith, and that he should be granted the proceeds of Wollaston's so-called Donation Fund, established to provide a modest income to a scientist whose career, through no fault of his own, was—if the double pun may be permitted—on the rocks.

William Smith was at Hackness when he was first given the momentous news, in a letter sent by hand from the society. His excitement and delight are infectious: One can sense the pleasure, the new spring in his step, the feeling of relief and gratitude and of a wrong being righted, *at last*!

"Last night I received Professor Sedgwick's official letter," he writes at the end of the month to his nephew—who himself was now well on in his own distinguished career.* "He presses me to appear in person at the Society's Anniversary on the 18th February," and where he would "go through the form" of being presented with the medal. "The form" was necessary because in fact the medal itself was not finished when time came for the award—but the society went ahead anyway, giving itself up to a state of what for the geological world is very close to rapture.

"At their meeting every countenance glowed with delight," Smith wrote to his niece,

> when the twenty guinea purse was delivered to me . . . then ninety merry philosophical faces glowed over a most sumptuous dinner at the Crown & Anchor. The new President Mr. Murchison took the Chair. On his right sat Mr. Herschel, Sir John Johnstone, Professor Sedgwick, myself, Mr. Blake, Dr. Fitton . . . after drinking much success to their fellow associates in science they drank my health, coupled with the numerous Geological Societies which now spot the range of the Oolitic series was given with three times three, which was truly drunk with enthusiasm.

Old William Smith sounds to have been almost a little tipsy while writing this letter; but disabled by drink or not, he had

*Phillips himself was awarded the Wollaston Medal in 1845. Buckland got it in 1848, Sedgwick in 1851, Fitton in 1852—and Charles Darwin, whose work on the origin of species owes much to geology, and a very great deal to Smith, won the Wollaston Medal in 1859. The one early geologist never considered was George Bellas Greenough.

clearly remembered to see to it that Sir John Johnstone had been invited, so that the man who had rescued him from rural obscurity was on hand to watch the revels of his elevation to the *corps d'élite* of the nation's scientists.

And as Adam Sedgwick was to make clear, it was a *corps d'élite* with but one member. The closing paragraphs of his short speech are worth quoting at their high-flown length, not least because they include the one phrase—an injudicious hyperbole, some churlish few will say—that has since marked the memory of William Smith out for all time, and that remains carved on his headstone and in gilt letters on a dozen public memorials.

> I for one can speak with gratitude of the practical lessons I have received from Mr. Smith; it was by tracking his footsteps, with his maps in my hand, through Wiltshire and the neighbouring counties, where he had trodden nearly thirty years before, that I first learnt the subdivisions of our oolitic series, and apprehended the meaning of those arbitrary and somewhat uncouth terms, which we derive from him as our master, which have long become engrafted into the conventional language of English geologists and, through their influence have been, in part, also adopted by the naturalists of the continent.
>
> After such a statement, gentlemen, I have a right to speak boldly, and to demand your approbation of the Council's award. . . . And if it be denied us to hope that a spirit like that of Wollaston should often be embodied on the earth, I would appeal to those intelligent men who form the strength and ornament of the Society, whether there was any place for doubt or hesitation? Whether we were not compelled, by every motive which the judgment can approve, and the heart can sanction, to perform this act of filial duty, before we thought of the claims of any other man, and to place our first honour on the brow of The Father of English Geology. . . .

> [I]t was he that gave the plan, and laid the foundations, and erected a portion of the solid walls, by the unassisted labour of his hands . . . .
>
> I think it a high privilege to fill this Chair, on an occasion when we are not met coldly to deliberate on the balance of conflicting claims in which, after all, we may go wrong, and give the prize to one man by injustice to another; but to perform a sacred duty where there is no room for doubt or error, and to record an act of public gratitude, in which the judgment and the feelings are united.

William Smith, now sixty-two years old, slightly lame from rheumatism, a little deaf, but otherwise as fit and wiry as a field geologist has a right to be, sat beaming throughout. He made a short speech of thanks, noting that Sir Isaac Newton had been born on the Oolite, and remarking on how the science of geology might have changed "had he looked down at the ground instead of up at the apple"—a remark that produced a clatter of (presumably polite) laughter that enabled Smith to resume his place and "hide my honoured head among the seated."

Before he did so, however, he presented the society with three documents that have remained in Burlington House apartments ever since. He had discovered among his papers the original manuscript version of the Table of Strata he had dictated to Benjamin Richardson, in Joseph Townsend's drawing room, thirty-two years before;* and he had found his circular map of Bath, colored in by hand in 1799; and he offered one of his elegant precursors to his great geological map of the nation shown as the frontispiece to the story.

There were more lunches and dinners the following day, an

*Townsend had died in 1816; but Benjamin Richardson was very much alive, and wrote from his vicarage at Farleigh Hungerford to say of Smith that he was "happy to hear the Geological Society proposes to pay a deserved compliment to his merits, to which I most gratefully bear a willing testimony."

artist was engaged to paint his portrait, and John Cary, who had published all the maps, atlases, and sections that were still on occasional sale in London, made Smith a present of a newly bought pair of silver-framed spectacles. And then Smith left London, heading first to Churchill, where he was born, and—despite never being much of a God-fearing man—going to church with all of his surviving family, where they engaged in "presenting our bald heads in supplication to their Maker."

A week later and he was back in the icy fastnesses of Yorkshire, reunited with his wife, who he reported to be in tolerable health though troubled by a head cold. The Father of English Geology was at last able to rest on his laurels and begin the final chapter of his life as a revered elder statesman, at long last accorded his due, and given the respect that his hitherto unsung achievements deserved.

EPILOGUE

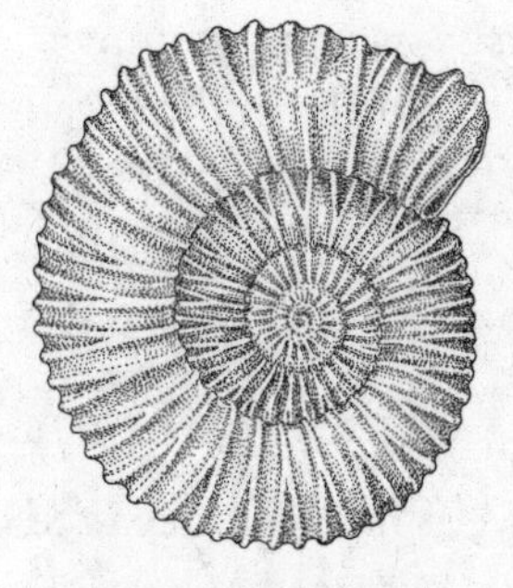

Pavlovia pallasioides

As everyone might wish, nothing but goodness seems to have attended the last years of William Smith's life. He moved from Hackness to his cottage in Scarborough, he cultivated a small garden, he read and tried (in vain) to write the story of his own life, he nursed his now-fast-fading wife, he went for long walks on the cliffs, he raised geese and sent them off to his relations, he wrote letters aplenty—letters that show him to be a happy old countryman, lost in the revelry of a comfortable old age.

A literary critic might have cruel amusement with his style. I have omitted from this account all of his innumerable attempts at writing poetry, the results of which were execrable at best. The paragraph that follows is not chosen for its merit as a piece of writing, but just to show the new-found contentment which had been so sorely missing from his life before. It was an autumn evening as he sat and wrote the following in his journal:

> When the sunbeam flickered on the gently waved gossamer suspended from the foliage, and filmy winged insects were waving up and down, innumerable, as if promiscuously enjoying the ethereal sports of a calm and sunny autumn morn, or that of one of summer's retiring days, I, after breathing from the Bridge Walk Terrace the pure air wafted off the ocean's smooth and boundless surface beneath its calm blue canopy—I so refreshed, and on my breakfast feasting—sat and enjoyed the same at my cottage window fronting that fine and full-grown hawthorn which casts a deep shade across the lawn and those bright distant lights which remind me of Claude's fine forest glades where flickering gleams of light seem to steal in between the trees. Even so complete is my seclusion and calm retreat from the busy town and its numerous far distant visitors who annual come to bathe in the ocean and drink at the Scarboro' Spa.
>
> Thus calmly to enjoy retirement with the never failing resources of a well-stored mind is the sweetest pleasure of a full-aged man.

He kept his "well-stored mind" as active as he could. He tried to tinker a little with the geology he knew, but steadily he came to realize that behind the vanguard of the "young Turks" who were now running the science, and who had given him his long-due recognition, geology was now accelerating rapidly away from him, the discoveries and theories multiplying at exponential rates.

He would travel, with near-religious punctiliousness, to the annual meetings of the newly formed British Association for the Advancement of Science—a body his nephew, John Phillips, had played a large part in establishing. But he would understand less and less of what was said; and when he came to suggest or to offer a paper, he found the organizers less than fully engaged by what he had to say. There were respectful pauses, hesitancies in their manner, suggestions about another day, another topic.

His papers, wrote the historian Hugh Torrens, in the entry on Smith in the *New DNB*, "demonstrated all too clearly Smith's great limitations in the new world of non-practical or theoretical geology, to which he was now expected to contribute." Slowly, though Smith probably never came fully to know it, and cruel though it seems to say so, he was becoming out-of-date. It seems fitting to suggest that, living so close to their playground in Yorkshire's Upper Jurassic, he was becoming a dinosaur himself in all but name.

But the comforts continued to arrive. In the early summer of 1832 Smith traveled to Oxford and in a further ceremony received his Wollaston Medal, now fresh from the die makers and engravers. The society had not stinted: the Wollaston had been made by no less a medal maker than Benjamin Wyon, who held the august post of Chief Engraver of the Seals and was responsible for designing and manufacturing some of the finest medals in the land, as well as various great seals with which nation-states—most of them colonial governments—impressed the final versions of their laws.

Oxford's Sheldonian Theatre, where Smith was awarded the first Wollaston Medal.

The medal was indeed made of gold; Smith's name was engraved on one side, between a beribboned twine of laurel and palm. Roderick Murchison presented it to him at a ceremony in the Sheldonian Theatre, the graceful seventeenth-century building that still represents the spiritual core of the great university. He made just a short speech, reminding the assembled gathering that they were in the presence of "a great original discoverer in English geology"; he suspended the medal, on its blue silk ribbon, around the old man's neck, and then stepped back to watch with pleased contentment as the assembled worthies of all England's sciences applauded a choice with which everyone seemed to agree. All Oxford, and all scientific Britain, were now giving William Smith their combined imprimatur.

The government, too, chipped in. Within days it was announced that, in consequence of a petition from a panel of scientists, King William IV had agreed to grant Smith a pension, for the remainder of his life, of one hundred pounds a year. It was thought to be only right and fair that a man "whose personal labours for the good of all," as it was put in countless petitions and other laudatory papers, had cost him his fortune—and, for a while, his freedom—should not be allowed to pass into old age a pauper. A hundred pounds was thought a sufficient sum to keep all wolves from doors; it was gratefully given, and the beneficence gratefully extended to cover the needs of his widow, when Smith himself died.*

He still had one further honor to receive and that, which was offered (so his nephew reported) "quite unexpectedly" during another British Association meeting, was an honorary doctorate

*In the event, Mary Ann Smith lived on for five years after her husband died. Her last days were spent in the York Asylum. Other than the fact that she suffered periods of insanity, there are conflicting views about her personality. One writer, Hugh Torrens, declared in the *New DNB* that she was "as unsuited for being the partner of a meditative philosopher as she could well be"; John Phillips himself said that Smith's own problems were exacerbated "by a still more severe and invincible torment—a mad, bad wife." But one of Phillips's friends who visited the asylum reported that "Mrs. S. was rather a favourite in the house, being generally very cheerful, and made the whole company laugh."

of letters. The association was meeting in Ireland; the provost and fellows of Trinity College, Dublin decided that Mr. Smith must be so elevated. The old man—by now sixty-seven, "his increasing deafness depriving him of a full share of enjoyments"—was according to John Phillips astonished, "and sufficiently alive to feelings more common in his youth" to be delighted with the title. The delight was general—though there was astonished dismay in some quarters, too, that the Father of *English* Geology had been honored not by one of *England's* academic institutions, but by the premier college in Ireland.

He dressed up in his new robes. He doffed his cap at the appropriate moment. He listened to the perorations in Latin. He mingled briefly with his fellow honorees—the three distinguished astronomers Sir Thomas Brisbane,* Professor Gerard Moll, and Francis Baily, and Louis Agassiz, the Swiss-American naturalist and fossil fish expert who had come up with the radical notion of *die Eiszeit*, the Ice Age. Whether, with his Oxfordshire accent, his deafness, his rheumatics, and his general innocence of urban sophistication, Smith felt comfortable in such august company, we can but wonder—John Phillips left a less than candid record of the occasion. In any case he was soon to be plucked away from the festivities, perhaps tactfully.

The dean of Trinity showed him around Dublin and took him up to the north of Ireland—which was then, unlike today, politically a part of the same country. The director of the Geological Survey of Ireland took him in hand and showed him around farms in the geologically fascinating parts of County Antrim. But though the tertiary basalts of the Giant's Causeway may well have enthralled him, his former reputation seems to have died hard in Ireland; before long, according to local accounts, the newly created Dr. Smith was busily answering questions from the farmers about the efficiency of their drains.

---

*After whom the Queensland capital was named.

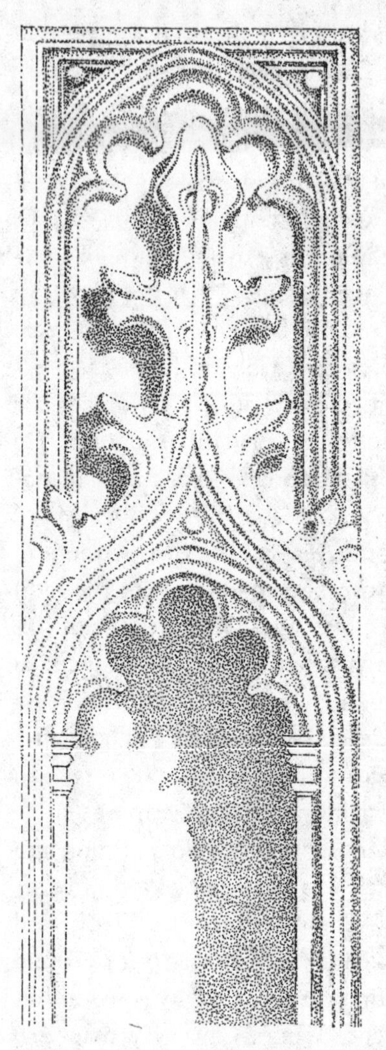

The ornate stonework of the House of Commons—originally a Permian limestone selected by a committee of which Smith was a member. It corroded badly; Charles Dickens inveighed against the choice.

To add to the final honor there was one final task. In 1834 the Palace of Westminster, where the British Houses of Parliament met, had been all but destroyed by fire. Sir Charles Barry had been commissioned, along with August Pugin, to conceive a fitting replacement—and duly did, with his son Edward Barry, creating what went on to become what is possibly the most familiar and iconic of all London's great buildings, crowned as it is with the clock tower that holds the famous bell, Big Ben. In 1838 a four-man committee was established—Barry, Henry de la Beche, an architect named Charles Smith, and, with his brand-new honorific, *Doctor* William Smith. The committee was charged with the express purpose of selecting the stone that was to be used in the construction of the giant new edifice.

The four men set out (by rail!—such were these modern times) from Newcastle upon Tyne after that year's British Association meeting.

They spent most of August and early September traveling by pony-and-trap, inspecting quarries, drawing up comparative tables of which stone was best for its workability, its appearance, its cost. The official report—to the commissioner of Her* Majesty's Woods, Forests, Land Revenues, Works and Buildings—was published in March the following year—five days before William Smith's seventieth birthday. It listed more than a hundred quarries; it compared the criteria of each; and it formally recommended that the new palace be built of a rock from whitish-colored Upper Permian formation, a sandy Magnesian limestone from a quarry at Mansfield Woodhouse, in Derbyshire.

It turned out to be a less than happy choice. The Mansfield quarry proved to have too little workable stone for the job, and another quarry at Bolsover had to be hastily substituted. Much the same happened there, with the stone running out within a matter of weeks. A third try, at a nearby quarry in the village of Anston, seemed to work, however. Before long a train of barges was setting out regularly along the Chesterfield Canal, to connect with the fast cargo sloops that sailed along the Trent and the Humber and thence down past the Wash and the Norfolk coast to the Thames Estuary and London. The quarry seemed bottomless. It provided, in the end, most of the stone for what were to be the visible upper parts of the palace for decades to come.

Except that the Anston stone—which is technically a part of the Cadeby formation of the Zechstein epoch of the Upper Permian—proved a dismal selection. The committee had clearly not imagined the havoc that would be wrought on smooth sheets of a corrodable, white-colored limestone by the action of the winds and rain of industrialized London. A foul cocktail of acids, smoke, sulfur, heavy metals, and all manner of gases quite unknown in the pleasant towns of North Yorkshire scoured the new buildings. The stone changed color, flaked away, peeled off,

*Queen Victoria had begun her reign the year before.

broke. All of a sudden the grandest public building in the land had the look of a dour tenement in the middle of a slum.

Within a decade of their completion, Barry's buildings were being roundly attacked for their appearance, for their ruined look, for their discoloration and decay. Charles Dickens led the fray: If the buildings of the day were going to be as ornate as Barry and Pugin seemed to want, he said, then let them be built of proper, durable stone. That chosen for the Houses of Parliament, Dickens fulminated in a newspaper article, "is the worst ever used in the metropolis."

And there were dark rumors about the stone selection committee—its members were accused, obliquely, of having merely junketed around the countryside, during a pleasant summer, at the taxpayers' expense, and not applying themselves properly to what was a profoundly serious business. It was Sir Henry de la Beche who took the brunt of public criticism. He had to testify before two select committees that were established specifically to make sure no public buildings ever suffered so embarrassing a fate again. No one seems to have blamed William Smith for anything: His involvement, said his admirers and detractors alike, was more of a sinecure for an honored ancient. Few had expected he would perform any kind of actual task on the committee, useful or otherwise.

✠

In the years since Barry and Pugin built their masterpiece, the Palace of Westminster has suffered heroically from the capital's pollution, and further committees have recommended all manner of replacement stones in the hope that some might cheat the acid rainfall. These days the Commons building has been wholly rebuilt, using another limestone entirely—the warm, honey-colored semioolitic limestone from the Middle Jurassic of Lincolnshire, known as Clipsham stone.

Smith would have found the ironies delectable: first that he managed to escape the opprobrium of the initial choice of a rock from the gloomy and fossil-free Permian; and second, that in the end it was his most favored horizon—the Jurassic oolitic formation for which he and his great map are best known—that was used to face his country's best-known public building. Once again, the allure of the lively and warm shallow seas of Middle England's Middle Jurassic proved inescapable and irresistible.

✢

William Smith set out for Birmingham for the next British Association meeting the following August. He had spent the previous week in Oxfordshire, wandering through a number of great gardens—which John Phillips points out were spread across basement rocks of both between the Jurassic and the Cretaceous. By this point Phillips seems to have become more than a little carried away by romantic notions about his elderly uncle's enduring interests—and has him placed here in summertime Oxfordshire, yet again pondering the great questions about Mesozoic stratification. One has to doubt it. More than likely this stooped, deaf, rheumatic old man was taking time just to smell the roses, and to keep out the possibility of a late summer chill.

This he did not manage. His coach stopped in Northampton, and as he was in no great hurry, he decided to put up for a few days with an old friend named George Baker. He caught a cold, suspected nothing, and went to bed. But the cold settled on his lungs, and it worsened. Doctors were called. John Phillips, already in Birmingham for the meeting, was summoned, and arrived in Northampton on the morning of Tuesday, August 27. There was little he could do. Uncomplaining or unknowing, William Smith died shortly after ten o'clock on the following night.

He was buried nearby, in Saint Peter's Church. There are a number of memorials—the most poignant, with the longest and most erudite eulogy, stands in Saint Peter's. It is a marble column, and it is topped, quite appropriately, with a copy of a bust by Matthew Noble. It was an appropriate choice indeed, since the sculptor was a man Smith had met and befriended in Hackness, the Yorkshire village where, eight years before, his life had changed so much for the better, and public recognition of his astonishing talents got properly under way.

Smith's own grave outside the church is topped by a massive block of sandstone, now so corroded that if there was any inscription, it is quite illegible. The original of Noble's bust of Smith stands in the museum in Oxford, near others of Buckland, Huxley, and Charles Darwin; there is another copy in the Geological Society, close to one of his nemesis, George Greenough. There is a tall monument standing beside the country bus shelter on the green in the little village of Churchill, where he was born; and a cul-de-sac of modern houses nearby has been named William Smith Close. There is a signpost pointing to Rugborne Farm, which Smith himself thought of as the birthplace of the science he created. There is a plaque on Joseph Townsend's house in Bath, where the table of the strata was dictated. There is another on the house at Tucking Mill—though as it happens, and typically for the memory of this often-forgotten, often-overlooked man, it is on the wrong house, and nobody has made any serious moves to try to shift it. And in the Geological Society there is a portrait. Beneath it, within the frame, there is a cut-out circle containing a lock of his hair, gray and curled, and apparently snipped off as he lay in the George Baker's upstairs bedroom on the evening he died.

Overlooking this vaguely macabre memorial is a long marble staircase, and high up on the wall beside it, a pair of pale blue curtains. Behind these curtains, if anyone can be found to pull the tasseled cord to part them, is the greatest of all memorials—the very map that William Smith created, all by himself, in a labor of love that lasted for fifteen years, or, in truth, for all his life.

Normally the map remains hidden from view, to keep it shielded from the light. The blue velvet curtains remain pulled firmly shut. The hundreds who pass by each day seldom glance up at it—and even if they do, few of them ever ask to see it, and fewer still ever stop to wonder who created it, and why, and exactly how.

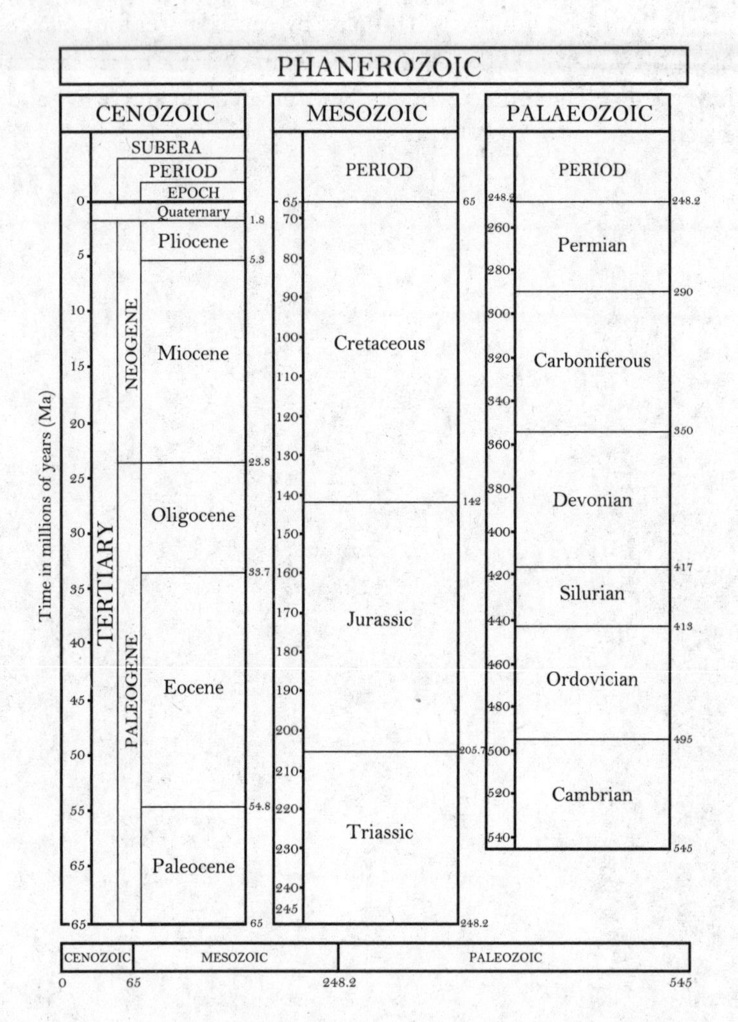

From F. M. Gradstein and J. Ogg, "A Phanerozoic Time Scale," *Episodes* 19, nos. 1 & 2 (1996).

Glossary of Geological and Other Unfamiliar Terms Found in This Book

Ammonite: A **mollusk**, with a shell coiled in on itself, containing folded internal compartment walls, living between the Devonian and Upper Cretaceous.

Annelid: A classic type of worm, having a long tubular body composed of a large number of rings.

Annularia: Needle-shaped leaves of the horsetail plant *Calamites* that are arranged in rosette-like whorls. (The leaf whorls are named separately from the stems, which are actually called *Calamites* and are a common fossil.

Anthracite: A hard, jet black, metallic-looking flammable **rock** that is the highest-ranking coal, the only version that can rightly be considered properly **metamorphic**.

Anthracosia: Nonmarine **bivalve** common in certian horizons in the British Carboniferous **Coal Measures**, and a useful stratigraphic indicator.

Arenaceous: Sandy, composed of sand grains.

Argillaceous: Rocks or **sediment**s that are composed chiefly of **clay** and clay minerals.

Arthropod: The largest phylum in the animal kingdom, with species that inhabit all kinds of environment: Typically they have segmented bodies, jointed appendages, and a hard carapace or exoskeleton. Insects and **trilobite**s are classic arthropods.

Ashlar: A hewn stone, cut into flat slabs, used for facing buildings.

Asteroceras: A classic Lower Jurassic **ammonite**, with an attractively ridged outer shell, common in the Lias of Yorkshire and elsewhere in northern England.

Batch: Waste pile from a Somerset coal mine

Bedding plane: The surface that divides one layer in a **sediment** from another, usually marked by a change in color, composition, or texture.

Belemnite: The hard, fossilized calcitic guard, usually pencil-shaped, that is often the only preserved part of a squidlike **cephalopod**.

Bituminous coal: The most common form of **coal**, softer and less pure than **anthracite.**

Bivalve: An aquatic **mollusk**, frequently fossilized, that has two calcareous shells that enclose the living parts of the animal.

Brachiopod: Solitary, bilaterally symmetrical **bivalve** marine invertebrate.

Brash: Loose broken rock forming the highest stratum beneath the soil of certain districts. Cornbrash, brashice. Perhaps a corrupt form of the French *brèche.*

Carboniferous limestone: Fossil-rich marine **limestone** laid down in the Carboniferous era, forming much of the characteristic landscape of the western Pennine Hills and northern Somerset.

Carstone: A brown **sandstone** with an interstitial cement of **limonite.**

Cephalopod: The most complex and highly organized class of **mollusks** (with a distinct tentacled head), which include the fossil **ammonites** and the modern squids, octopuses, and pearly nautilus.

Cirripede: A low class of **crustacean**, such as a barnacle.

Chalk: A soft, porous, and fine-grained **limestone,** characteristically white in color, most memorably found in the cliffs of southern Kent—"the white cliffs of Dover."

Chert: A form of quartz, very finely crystalline, that often forms irregular nodules in other **sediments**.

Chronostratigraphy: A geological time scale that uses fixed and internationally defined standard reference points.

Clay: A **sediment** defined as having particles less than four microns in diameter.

Clunch Clay: William Smith's somewhat inelegant—and quickly replaced—name for what is now called the Oxford Clay.

Coal: A combustible organic **sedimentary** rock formed of compressed decomposed plant remains.

Coal measures: A Western European **stratigraphic** term used for the **rocks** of the Upper Carboniferous.

Coelacanth: A type of bony fish thought to have been extinct for fifty million years before the discovery of a living specimen, classified as *Latimeria chalumnae*, off the South African coast in 1938.

Coprolite: Fossilized animal droppings.

Coral Rag: Coarse-grained rubbly limestone—one of the oldest building stones used in Oxford and the surrounding villages. William Smith's name for the formation now included in the Corallian.

Cornbrash: William Smith's formation name, taken from Wiltshire dialect, for a coarse, shelly limestone of the Great Oolite that disintegrates and breaks up easily, providing soils ideal for growing corn. See also **brash**.

Crinoid: A class of **echinoid** comprising the stalked sea lilies and feather stars.

Crust: The outermost solid layer of the earth, separated from the underlying **mantle** by the so-called Mohorovicic Discontinuity.

Cycad: One of the more important fossil floras of the Mesozoic, resembling small palm trees.

Cyclothem: A recurring sequence of beds found in many **sedimentary rocks**, notably in the Upper Carboniferous, in coal-rich series.

Dip: The angle in degrees between the horizontal and any inclined geological feature, most usually the **bedding plane** of a series of **sediments**.

Dislocations: Any surface across which there is a loss of continuity—for example, a **sediment** interrupted by a **fault**.

Drift: Superficial geological material in a landscape, often brought by ice or glacial meltwater. The term is used in geological mapping, and is distinguished from **solid**.

Echinoid: The class of **echinoid** to which sea urchins and sand dollars belong.

Estwing: Noted American manufacturer, based in Rockford, Illinois, of geological and other hammers.

Facies: The combined **lithological** and **paleontological** characteristics of a **sediment** from which it is usually possible to infer the environmental conditions at the time of deposition.

Fault: A fracture in a geological structure, caused by movement and shear across a surface, resulting in displacement.

Flint: A microcrystalline form of silica found usually in **chalk**, which in other **sediments** is called **chert.**

Fool's gold: Iron pyrite, usually in cubic crystals that have the appearance of yellow metal.

Forest Marble: An argillaceous laminated shelly limestone in the Great Oolite Series, which when polished passes as marble. Named by William Smith after the Forest of Wychwood, which in his time extended over a much greater area.

Fossil: Originally an adjective meaning "ancient"; now a noun used to denote the preserved remains or evidence, usually found in **rocks**, of previous organic life.

Friable: Property of a **rock** that can be disintegrated by the slightest pressure, such as that between two fingers.

Freestone: A fine-grained **rock** that can be cut and sawn with little risk of fracturing.

Fuller's Earth: Term used either for the **montmorillonite** clay employed in the wool industry to remove lanolin or in stratigraphy to define an upper part of the Middle Jurassic in southern England.

Gastropod: A **mollusk** with a single coiled shell and a foot, the anterior part of which is developed into a head.

Genus: A **taxonomic** term, low in rank (between Family and Species), and usually consisting of closely related species.

Geology: The study of the solid part of a planet.

Geotherm: The variation of a temperature with depth in the earth.

Gondwana: Former southern supercontinent.

Goniometer: An instrument for measuring the angles between the faces of crystals.

Graptolite: A tubelike creature, with branched extensions, found in many **shales** and **slates** of Ordovician and Silurian age.

Gymnosperm: Plants—frequently found fossilized in Carboniferous **rocks**—with their seeds held in cones, as with cycads and ferns.

Hercynian: A term still occasionally in use for a Permo-Triassic **orogenic** episode; generally called the **Variscan.**

Horsetail: A type of fern commonly found in Carboniferous **rocks**.

Ichthyosaur: A fishlike marine reptile, becoming extinct in the Middle Cretaceous.

Igneous: One of three primary **rock** types found in Earth's **crust** (see also **metamorphic**, **sedimentary**), igneous **rocks** have solidified from molten **magma** generated deep within the planet.

Iron pyrite: A metallic-looking sulfide of iron, known jocularly as **fool's gold.**

Kellaway's Beds: Richly fossiliferous calcareous clay and sandstone beds forming the basement of the Oxford Clay, and resting directly on the **cornbrash**. Smith's term for it was "Kelloway's Stone."

Laurasia: Former northern supercontinent.

Lepidodendron: A tree, commonly found fossilized in Carboniferous **coal** deposits, notable for its size—up to one hundred feet tall—and scaly bark.

Lias: Somerset quarryman's term, from either the pronunciation of the word "layers" or the French for a kind of **limestone**, which now applies to a period of the Lower Jurassic.

Lignite: Low-grade brown-black **coal**.

Limestone: A **sedimentary rock** of which the principal ingredient, from either shelly remains or precipitates, is calcium carbonate.

Limonite: A yellow-brown inferior iron ore, often found in large earthy masses in a variety of forms.

Lingula: A type of **brachiopod**, still extant, but known from as far back as the Cambrian.

Lithology: A description of the major macroscopic features of a type of **rock**—particularly texture, color, and composition.

Magma: A molten fluid of complex silicates and metals formed within the **crust** or upper **mantle** of the earth, which may be extruded by volcanic eruptions or other **igneous** activity to form solid and consolidated rocks.

Magnesian limestone: A **limestone** in which some ten percent of the calcium carbonate has been replaced by a magnesium carbonate, resulting in a much denser and harder **lithology**.

Mantle: The zone in the earth between the hard crust and the liquid metallic core.

Marlstone: A half-formed **argillaceous limestone**, with a large proportion of **clay**, microfossils, and relic ooze.

Metamorphic: One of the three basic **rock** types (see also **sedimentary** and **igneous**) in which pressure, temperature, and time have greatly deformed and recrystallized the original.

Millstone Grit: A coarse **sandstone**, found in the Carboniferous of Britain, much used for building in the English North.

Mollusk: A phylum of animals with a fleshy body usually surrounded by a calcareous shell or shells.

Montmorillonite: See also **smectite**. A sheet silicate mineral, found in some **clays** (most notably **fuller's earth**) and widely used for its absorptive powers.

Mosaic theory: Used to describe the general belief among early geologists in divine creation of the earth, and in the occurrence of the **Noachian** flood, or Deluge.

Neptunism: A theory, popularized by **Werner**, which held that all **rocks** were precipitates from a primordial ocean.

Noachian: Pertaining to Noah, and to the persistent belief among nineteenth century geologists that the Flood with which he was associated had profound and global geological significance.

Oolite: A common **limestone** composed primarily of tiny spherical accretions of calcium carbonate around a quartz core.

Orogeny: A period of **crustal** compression that results in mountain-building and consequent major changes in geological conditions.

Ostrea: The oyster.

Outcrop: That part of a **rock** unit that is exposed at the earth's surface.

Overburden: Any loose material that overlies bedrock.

Paleogeography: The reconstruction, inferred from geological evidence, of the former physical geography of the earth.

Paleontology: The study of fossil flora and fauna, in the attempt to infer both the age and relative age of the **sediments** in which the fossils appear, and gauge the nature of the contemporaneous environment.

Pangea: A supercontinent that existed (according to most theories) for some forty million years from the late Permian to the Triassic. It was, according to these same theorists, surrounded by a superocean, the Panthalassa.

Pantograph: A jointed parallelogram of slender rods, used by surveyors for copying diagrams and plans on the same (or on a different scale).

Peat: A thick organic soil deposit that, if dried, is flammable, burning with a characteristic sweet smell. In some senses regarded as a very primitive form of **lignite,** itself a low-grade **coal**.

Petrifaction: One process of making a fossil, in which chemical precipitates are deposited within the porous structure of the shell or other hard parts of an organism.

Phyllosilicate: A group of silicates with crystal structures arranged in large sheets, common in many **argillaceous** and low-grade **metamorphic rocks**.

Phylum: The second highest **taxonomic** classification, one of the main divisions of animal and plant kingdoms.

Plane table: A surveyor's instrument, used in conjunction with a sighting glass, for marking the relative positions of observed structures and the angles between them.

Plesiosaur: A large aquatic reptile, in appearance like (as William Buckland put it) "a snake strung through a turtle," common in the Jurassic.

Plutonism: The theory, advanced by Hutton, that held that almost all **rocks** originated as a result of heat and melting, rose from the **mantle** to form new land, only to decay and be regenerated.

Portland Stone: An Upper Jurassic **mollusk**-rich **limestone**, commonly used for building the grander structures and monument in the large British cities.

Pterodactyl: A **fossil** reptile found in Jurassic and Cretaceous **sediments**, noted for its large jaw and membrane attached to the long fourth digit of its forelimb, enabling it to fly.

Quarry sap: The trace liquids found in some **freestones**, which, on freezing, allow the **rock** to be split along its **bedding planes**.

Radiometry: A common means of determining the age of **rocks** by measuring the relative amounts of "parent" and "daughter" isotopes caused in radioactive decay, the most common pairings being potassium-argon, rubidium-strontium, and samarium-neodymium

Red Marl: Smith's name for the Keuper Marl, which consists of red fine-grained siltstones forming the upper part of the Trias.

Rock: A consolidated or unconsolidated aggregation of mineral or organic matter, formed either by the accretion (**sedimentary rocks**) of grains or sediments, by the crystallization (**igneous rocks**) of molten material, or by the alteration (**metamorphic rocks**) of existing **rocks** under pressure and heat.

Sandstone: A **sedimentary rock** composed of sand-size particles of silicates bound together by a cement that may be carbonate

Seat earth: A **fossil** soil, often with plant rootlets still in place, often found immediately beneath a layer of **coal.**

Sediment: Solid material—organic and inorganic—that has settled from suspension in a liquid, usually water.

Sedimentary: Consolidated **sediments**, usually with organized into **strata** or with **bedding** characteristics.

Seismic: Concerned with the vibration of the earth, whether naturally or artificially induced. Pertaining to earthquakes and crustal movement.

Shale: An **argillaceous rock,** noted for its thin and well-defined laminations.

Siltstone: A consolidated silt, or a **clay**like **rock** with particles measuring larger than four microns.

Slate: A low-grade **metamorphic** rock, usually **argillaceous**, and that, because of the extremes of pressure and temperature to which it has been subjected, has well-defined cleavage and **bedding planes**.

Smectite: The **montmorillonite** group of clay minerals, useful for leaching oil from wool.

Solid: Term used in geological cartography to describe the bedrock, and opposed to the superficial and often glacially derived material, known as **drift.**

Stratification: The layered or bedded arrangement of **rocks**, usually but not uniquely found in **sedimentary** deposits. There can be stratified lava flows and **metamorphic rocks.**

Stratum (pl. strata): A defined layer of **sedimentary rock**, usually separated from other beds above and below by **bedding planes.**

Striations: Small marks and lines, frequently parallel, etched into a solid surface by some external—often glacial—force.

Strike: The direction taken by a structural surface—most usually a **bedding plane**, and also a **fault**—as it intersects the horizontal. Strike is at ninety degrees to the direction of **dip**.

Stromatolite: A cumbersome fossiliferous mass of an evidently funguslike former nature.

Taxonomy: The classification of plants of animals. The main taxa, in descending order, are kingdom, phylum, class, order, family, genus, and species.

Tectonic: Used in reference to the formation of a major earth structure, usually involving deformation or collision.

Terebratulid: Order of small brachiopods first appearing in the Devonian, common in the Jurassic, and known in English rural areas as lamp shells.

Tertiary: One of the great and more recent divisions of the geological time scale, which includes the Eocene, Oligocene, Miocene and Pliocene; hence *Tertiaries*, rocks laid down or created—like *Tertiary basalts*—during this period.

Tethys: The former ocean that broadly separated the two great Mesozoic supercontinental landmasses of **Gondwana** and **Laurasia.**

Theodolite: A surveyor's instrument for measuring vertical and horizontal angles.

Titanites: A very large and classical **ammonite.**

Trilobite: A common **arthropod,** with nearly 4000 species, found from the Cambrian to the Permian. Its biography, notable for its importance in displaying Darwinian evolutionary principles, was written by R. Fortey.

Unconformity: Surface of contact between two series of **rocks** that is sufficiently unconformable to imply a passage of time and often the occurrence of earth movement between the two periods of deposition.

Uniformitarianism: James Hutton's belief, formulated in the late eighteenth century, that the natural occurrences observable today have been occurring for millions of years past, and thus—as in "the present is the key to the past"—indicate the basic processes of **geology.**

Variscan: A major orogeny, occurring during the Carboniferous and Permian, relating to the closure of the gap between Africa and Europe. It resulted in the building of many central European mountain chains. See also **Hercynian.**

Vein: A deposit of a mineral, usually crystalline, limited to a fissure or joint of a **rock;** to be compared with a lode, which involves a much wider dissemination of a mineral through a rock body.

Wernerian: The principles outlined by Abraham Werner of Freiburg, Germany, in the early nineteenth century, which were broadly based on **Neptunism,** and inspired a large school of mapmakers and students of **geology**—which Werner himself preferred to term "geognosy."

Wollastonite: A silicate mineral noted for its long fibers, used in the making of the insulating material known as rock wool; it was named after William Wollaston, the medal in whose name remains the highest honor that can be given a geologist. The first winner, in 1831, was William Smith.

Sources and Recommended Reading

The papers, diaries, sketches, execrable poems, and extraordinary maps of William Smith are kept in the archive of the University Museum, Oxford, as are the papers of his nephew and future Oxford professor of geology, John Phillips, and those of the flamboyantly eccentric omnivore Dean William Buckland. The collections of George Bellas Greenough are in the archives of the Geological Society of London. There are other important papers housed in the Eyles Collection at the University of Bristol.

A very few of the books that are listed below will make enjoyable reading—most notably the two enlightening works by the eminent paleontologist Richard Fortey, *The Hidden Landscape* and *Trilobite!*; Noel Annan's highly readable study of Dean Buckland in *The Dons*; and Roger Osborne's most original *The Floating Egg*. Other books that seem likely to appeal to the general reader I have marked with an asterisk.

The greatest of all the works noted here—aside, of course, from Darwin—is the majestic tome (no other word can possibly do justice) written in 1933 by W. J. Arkell: *The Jurassic System in Great Britain*. This utterly beautiful book, elegant in design and writing, represents the life's work of a man who was passionately fascinated by the most celebrated—and, one might say, looking at the rocks and villages along its outcrop, the most *English*—of all the geological periods. It has long been out of print, and a clean copy will cost a good deal of money. But to anyone whose interest in geology at its best may have been piqued by this short account, I urge them—find yourself an Arkell, buy it, and treasure for yourself and for your descendants. There are all too few books of its like.

✠

To write this book I made use of the following:

Allaby, A., and M. Allaby. *The Oxford Dictionary of Earth Sciences.* Oxford, England: Oxford University Press, 1999.

Allsop, Niall. *The Somersetshire Coal Canal Rediscovered.* Bath, England: Millstream Books, 1993.

*Annan, Noel. *The Dons.* London: HarperCollins, 1999.

*Arkell, W. J. *The Jurassic System in Great Britain.* Oxford, England: Clarendon Press, 1933.

Bassett, Michael G. "Formed Stones." *Folklore and Fossils.* Cardiff: National Museum of Wales, 1982.

Bennett, Stewart. *A History of Lincolnshire.* Chichester, England: Phillimore & Co., 1999.

Berger, Lee. *In the Footsteps of Eve.* New York: Simon & Schuster, 2000.

*Bernal, J. D. *Science in History.* London: Watts & Co., 1954.

Blundell, D. J., and A. C. Scot. *Lyell: The Past Is the Key to the Present.* London: Geological Society of London, 1998.

*Briggs, Asa. *A Social History of England.* London: Penguin, 1987.

Brooke, J., and G. Cantor. *Reconstructing Nature.* London: T. & T. Clarke, 1998.

Brooke, John Hedley. *Science and Religion.* Cambridge, England: Cambridge University Press, 1991.

Brown, Roger Lee. *A History of the Fleet Prison, London.* Lampeter, England: Edwin Mellen Press, 1996.

Clew, Kenneth. *The Somersetshire Coal Canal and Railways.* Newton Abbott, England: David & Charles, 1970.

Cox, L. R. "New Light on William Smith and His Work." *Proceedings of the Yorkshire Geological Society* 25, pt. 1, 1942.

———. *William Smith and the Birth of Stratigraphy.* International Geological Congress, 1948.

Craig, G. Y. *The Geology of Scotland.* London: Geological Society of London, 1991.

Craig, G. Y., and J. H. Hull. *James Hutton—Present and Future.* London: Geological Society of London, 1999.

*Darwin, Charles. *The Origin of Species.* New York: New American Library, 1958.

Daunton, M. J. *Progress and Poverty.* Oxford, England: Oxford University Press, 1995.

Davies, G. L. "The University of Dublin and Two Pioneers of English Geology." *Hermathena* 109 (1969).

Doyle, Peter. *Understanding Fossils.* New York: Wiley, 1997.

Doyle, Peter, and Matthew Bennett, eds. *Unlocking the Stratigraphical Record.* New York: Wiley, 1998.

Duff, P. McL. D., and A. J. Smith, eds. *The Geology of England and Wales.* London: Geological Society of London, 1992.

Eastwood, T. *Stanford's Geological Atlas.* London: Edward Stanford Ltd., 1964.

Edmonds, J. M. "The Geological Lecture-Courses given in Yorkshire by William Smith and John Phillips, 1824–1825." *Proceedings of the Yorkshire Geological Society,* 1975.

———. "The First 'Apprenticed' Geologist." *Wiltshire Archaeological and Natural History Magazine* 76 (1981).

Eldredge, Niles. *The Triumph of Evolution and the Failure of Creationism.* San Francisco: W. H. Freeman & Co., 2000.

Emsley, Clive. *Crime and Society in England, 1750–1900.* London: Longman, 1987.

Eyles, Joan M. "William Smith: The Sale of His Geological Collection to the British Museum," *Annals of Science,* 23, no. 3 (1967).

———. "William Smith (1769–1839)—a Bibliography of his Published Writings, Maps and Geological Sections, Printed and Lithographed." *Journal of the Society for the Bibliography of Natural History.* (April 1969).

———. "William Smith: Some Aspects of his Life and Work." In C. J. Schneer, ed. *Towards a History of Geology.* Cambridge, Mass.: MIT Press, 1969.

———. "William Smith, Richard Trevithick and Samuel Homfray: Their Correspondence on Steam Engines 1804–1806." *Transactions of the Newcomen Society* 43 (1970–71).

———. "William Smith's Home Near Bath: The Real Tucking Mill." *Journal of the Society for the Bibliography of Natural History* (1974).

———. "G. B Greenough, FRS (1778–1855)." *Nature,* April 16, 1955.

Fearnsides, W. G., and O. M. B. Bulman. *Geology in the Service of Man.* London: Pelican, 1944.

*Fortey, Richard. *The Hidden Landscape.* London: Pimlico, 1993.

*———. *Trilobite!* London: HarperCollins, 2000.

Geikie, Sir Archibald. *The Founders of Geology.* London: Macmillan, 1897.

Gillispie, Charles Coulston. *Genesis and Geology.* Cambridge, Mass.: Harvard University Press, 1996.

Gould, S. *The Lying Stones of Marrakech.* New York: Harmony Books, 2000.

*———. *Wonderful Life.* London: Penguin, 1989.

Grantham, John. *The Regulated Pasture—a History of Common Land in Chipping Norton.* Chipping Norton, England: J. Grantham, 1997.

Green, G. W. *British Regional Geology: Bristol and Gloucester Region.* London: Her Majesty's Stationery Office, 1992.

Greene, John C. *The Death of Adam.* Ames: Iowa State University Press, 1959.

Hains, B. A., and A. Horton. *A British Regional Geology: Central England.* London: Her Majesty's Stationery Office, 1969.

Hardy, Peter. *The Geology of Somerset.* Bradford on Avon, England: Ex Libris Press, 1999.

Harland, W. B., et al. *A Geological Time Scale, 1989.* Cambridge, England: Cambridge University Press, 1990.

*Hawkes, Jacquetta. *A Land.* London: Cresset Press, 1953.

Hill, Christopher. *Reformation to Industrial Revolution.* London: Penguin, 1992.

*Holmes, Arthur. *Principles of Physical Geology.* New York: Nelson Thornes, 1993.

Hutton, James. *Theory of the Earth, Vol. III* (facsimile). London: Geological Society of London, 1997.

Innes, Joanna. "The King's Bench Prison in the Later Eighteenth Century." In John Brewer and John Styles, eds., *An Ungovernable People: The English and Their Law in the Seventeenth and Eighteenth Centuries.* London: Hutchinson, 1980.

*Jones, Steve. *Darwin's Ghost.* New York: Random House, 2000.

Kearey, Philip. *The New Penguin Dictionary of Geology.* London: Penguin, 1996.

Knell, Simon J. *The Culture of English Geology.* Aldershot, England: Ashgate Publishing, 2000.

Korsmeyer, Jerry. *Evolution & Eden.* Mahwah, N.J.: Paulist Press, 1998.

Lapidus, Dorothy F., with I. Winstanley. *The Collins Dictionary of Geology.* London: HarperCollins, 1990.

Laudan, Rachel. *From Mineralogy to Geology: The Foundations of a Science.* Chicago: University of Chicago Press, 1987.

Le Bas, M. J., ed. *Milestones in Geology.* London: Geological Society of London, 1995.

Lindberg, David, and Ronald L. Numbers, eds. *God and Nature.* University of California Press, 1986.

Lyell, Charles. *Principles of Geology.* London: John Murray, 1834.

McClay, Keith. *The Mapping of Geological Structures.* New York: Wiley, 1987.

McKibben, Bill. *The End of Nature.* New York: Doubleday, 1999.

Mather, Kirtley. *Source Book in Geology.* Cambridge, Mass.: Harvard University Press, 1967.

Meades, Eileen. *The History of Chipping Norton.* Chipping Norton, England: Bodkin Books, 1984.

Melville, R. V. and E. C. Freshney. *British Regional Geology: The Hampshire Basin.* Her Majesty's Stationery Office, 1982.

Numbers, Ronald L. *Creation by Natural Law.* Seattle: University of Washington Press, 1977.

———. *The Creationists.* Berkeley: University of California Press, 1992

———. *Darwinism Comes to America.* Cambridge, Mass.: Harvard University Press, 1998.

*Osborne, Roger. *The Floating Egg.* London: Pimlico, 1999.

Packard, Lisa. *Dr. Johnson's London.* London: Weidenfeld & Nicolson, 2000.

Pevsner, Nicolaus, and John Harris. *The Buildings of England: Lincolnshire.* London: Penguin, 1964.

Phillips, John. "Biographical Notice of William Smith, LLD." *Magazine of Natural History* (1839): 213.

———. *Memoirs of William Smith, LLD.* London: John Murray, 1844.

Plumb, J. H. *England in the Eighteenth Century.* London: Penguin, 1990.

Porter, Roy. *The Making of Geology.* Cambridge, England: Cambridge University Press, 1977.

Priestley, Philip. *Victorian Prison Lives.* London: Pimlico, 1999.

Robson, Douglas A. *Pioneers of Geology.* Newcastle-upon-Tyne, England: Natural History Society of Northumberland, 1986.

Rudwick, Martin. *The Great Devonian Controversy.* Chicago: University of Chicago Press, 1985.

Rule, John. *The Laboring Classes in Early Industrial England, 1750–1850.* London: Longman, 1986.

Rupke, Nicolaas. *The Great Chain of History.* Oxford. England: Clarendon Press, 1983.

Sale, Richard. *A Guide to the Cotswold Way.* Marlborough: Cordwood Press, 1999.

Serest, Michel, ed. *A History of Scientific Thought.* Oxford: Blackwell, 1995.

Sheppard, Thomas. *William Smith: His Maps and Memoirs.* Hull, England: Brown & Sons, 1920.

Singer, Peter. *A Darwinian Left.* New Haven, Conn.: Yale University Press, 1999.

Smith, E. A. *George IV.* New Haven, Conn.: Yale University Press, 1999.

Smith, Peter L. *Canal Architecture.* Princes Risborough, England: Shire Publications, 1997.

Stanforth, Alan. *Geology of the North York Moors*. National Park Information Service, 1993.

Strahler, Arthur M. *Science and Earth History: The Evolution/Creation Controversy*. Buffalo, N.Y.: Prometheus Books, 1987.

Thomson, David. *England in the Nineteenth Century*. London: Pelican, 1991.

*Toghill, Peter. *The Geology of Britain*. Shrewsbury, England: Swan Hill Press, 2000.

Tonga, Neil, and Michael Quincy. *British Social and Economic History, 1800–1900*. London: Macmillan, 1980.

Torrens, H. S. "Early Maps of the Somersetshire Coal Canal." *Cartographic Journal* (June 1974).

Truman, A. E. *Geology and Scenery*. London: Pelican, 1949.

Very, David, and Alan Brooks. *The Buildings of England: Gloucestershire 1*. London: Penguin, 1999.

Vile, Nigel. *Exploring the Kennet & Avon Canal*. Newbury, England: Countryside Books, 1992.

Watkins, Alan. *Churchill and Sarsden*. Gloucester, England: Alan Sutton Publishing, 1988.

Wicander, Reed, and James Monroe. *Historical Geology*. Pacific Grove, Calif.: Brooks/Cole, 2000.

Woodward, Horace. *The History of the Geological Society of London*. London: Geological Society, 1907.

Ziegler, Peter. *Geological Atlas of Western and Central Europe*. London: Shell International Petroleum BV, 1990.

Acknowledgments

In writing this book I owe the very greatest debt to Professor Hugh Torrens, the renowned historian of geology latterly based at Keele University in the English Midlands, who probably knows more about William Smith than anyone else alive, and is indeed himself in the process of writing the definitive academic study of Smith's career and legacy. He gave generously of his time, his advice, and his help, and handed me an immense number of his own most useful papers, both published and unpublished, from which I learned much; a lesser man, on learning that a rival biography was in the making, would certainly have reacted more coolly. I thank Professor Torrens for his magnanimity, and can only repeat now what I suggested to him at the time—that this short book should be thought of simply as the hors d'oeuvre while we wait in eager anticipation for his main dish, soon to come. I earnestly hope that he will find that this brief account—while not so scholarly as the work he plans—will be a worthy enough tribute to the shadowy and half-forgotten figure whom we both so much admire. I wish also to record my thanks to the tireless and indefatigable Soun Vannithone, who, though taking no time off at all from cooking his legendary Laotian cuisine at a pub (the Racing Page, in Richmond, well worth the detour), managed to complete, precisely on time, the intricate and delicate illustrations on these pages. Alan Davidson, who since his time as British ambassador in Vientiane, has kept in close touch with Soun, helped at all stages; and to Alan and Jane Davidson I offer my sincere gratitude.

Professor Jim Kennedy, at the University Museum, Oxford, kindly made available the papers of William Smith, William Buckland and John Phillips; Stella Brecknell, the librarian and archivist who oversees this magnificent treasure-trove of documents, proved of enormous assistance, seeing to it that almost all of the most interesting items in her care made their way to me on the remote Scottish island where, perhaps perversely, I chose to write this book. Professor Keith Thomson, also at the University Museum, gave me helpful advice about the impact on pre-Darwinian thought that came about as a result of William Smith's discoveries and theories.

Many of those who lectured in geology at Oxford when I was an undergraduate there in the mid-sixties remain in what is now grandly called the Department of Earth Sciences, and were each in their own way keen to help their prodigal student who, after so long a time away, decided to stumble back into writing about their discipline. In particular I wish to thank David Bell, Steve Moorbath, and Stuart McKerrow, whose lectures on igneous geology, geochemical dating techniques, and Jurassic paleontology respectively evidently left more of an impression on me than my generally lackluster examination results suggested.

Ron Oxburgh—now the Lord Oxburgh—was also at the Department in the sixties, and lectured on structural geology: he too has been helpful in more ways than the simply technical, not least because of his presidency of the Geological Society of London: I have many reasons for wishing to offer my gratitude for his efforts and enthusiastic support of this project. Rachel Laudan, from her home in Mexico, wrote helpfully about her own early interest in William Smith, and kindly sent me her entire doctoral thesis and several other papers that threw new light on Smith's many achievements. That her position has long been generally critical of Smith did not in the least diminish her support for this book: the fact, she wrote, that he had mapped all England, and essentially on his own, has long since persuaded her that Smith was indeed a remarkable man, and she has long thought he deserved a biography—providing only that it stopped short of suggesting that he deserved a sainthood. I hope that in this I have been temperate, and fair.

For various specific items of help and advice I wish also to thank: Robin Cocks, Jill Darrell, Richard Fortey, Ann Lum, Susan Snell and Brian Rosen at London's Natural History Museum; Wendy Cawthorne at the Geological Society of London; David Buchanan of the Scarborough City Museums; my friend Francis Herbert at the Royal Geographical Society; my long-term traveling companion Kirk Johnson at the Denver Natural History Museum in Colorado; his colleague there, Bob Raynolds; Ian MacGregor of the Meteorological Office Archive, who seems to be able to find out what the weather was like on any particular day in the last three centuries; the authors Simon Knell and Roger Osborne, who have both written fascinating recent books on the development of geological thought; Nicolaas Rupke, who is an academic specialist in Holland researching this the same, very English field of study; Patrick Wyse-Jackson, the geology archivist at Trinity College, Dublin; Robert Millspaugh of the American Association of Petroleum Geologists; Professor Ronald Numbers at the University of Wisconsin in Madison, for his views on evolutionists; Joanna Innes of Somerville College, Oxford, a specialist on early London prisons, for her help with details of life in the King's Bench debtors' prison; Derek and Eileen Brown for their hospitality and friendship in Chipping Norton, Oxfordshire, close to where William Smith was born; Brian Excell and Fiona Ann Drury for their comments on the Tisbury Coral; Denys Brunsden, winner of a William Smith Award, for his help on the Jurassic of Dorset; Lord and Lady Derwent of Hackness Hall, near Scarborough, for their hospitality and help when I arrived to ask about William Smith's Yorkshire exile; and Heather MacFadyen of Bristol, who kindly searched, with great professional expertise, the famous collection of the late Victor and Joan Eyles, a couple who—because of their profound knowledge of the subject—should by rights have long ago written the book I am writing now. My hope is that they would approve of the work I have done in their stead. My son Rupert Winchester also searched the papers in the Public Records Office at Kew—under the invigilation of a supremely helpful staff, he says—for details of Smith's imprisonment, and less fruitfully, his marriage. Juliet Walker was tirelessly helpful, as she has been for so many of my projects: I hope she finds the Aeron chair I sent from Oklahoma at least a comfortable small recompense for all she managed to do.

My editors in London, Anya Waddington, Juliet Annan, and Clara Farmer, have proved wonderfully sympathetic in dealing with the complications inherent in a book about so curious a subject as geology—with Clara especially so since her father, David Farmer, is a geologist and very kindly looked over his daughter's shoulder to make sure there were not too many errors of fact or judgment. Donna Poppy in London and Sue Llewellyn in New York, each deploying her remarkable copyediting skills, helped make sure that such infelicities as remained were ironed out and smoothed away. The proofreading phase of this book happened to coincide with my brief stay as a visiting professor at the University of Chicago, where I decided to take advantage of my situation by asking the members of my writing class if they would each care to look at a couple of chapters to try to spot the most egregious of errors. They managed to detect some;

and so I am happy to record my thanks to a group of clever and talented young men and women who I suspect—since most of them hope to become writers—will become distinguished and familiar bylines before very long. The names for which editors should thus be on the lookout are those of Amy Biegelsen, Robert Peter Cuthbert, Melissa Klimala Dean, Gina DiPonio, Kristen Ina Grimes, Kurt Hagstrom, Frank Karabetsos, Daniel Lavetter, Kathleen Lingo, Zachary Martin, Kristen Morgan Miller, Casey Sanchez, Vanessa E. Raizberg and Leslie Synn. Responsibility for those mistakes that managed to survive their scrutiny—and I hope there are few—should be laid squarely at my door alone. In New York the legendary Larry Ashmead—who, by extraordinary chance, was once a geologist too, but moved on to become one of the most cherished editors in American publishing—seemed to think the manuscript passed muster, and made criticisms that were as constructive as they could only be, coming from a publisher who knew his rocks. My agents—Peter Matson in New York and Bill Hamilton in London—were also enthusiastic about my telling the tale of William Smith, L.L.D.: I hope they will think the finished product lives up to their own expectations, which they communicated with such early eagerness to the publishers.

I wish finally to make mention of my unforgettable tutor at Oxford, the great stratigrapher, field-trip speed-walker, longtime supporter, and friend Harold Reading, who over three long years hammered geology into my head with about the same energy that, in the field, he hammered fossils out of limestones. Harold succeeded, if not in winning me the greatest of all degrees, nor in persuading me to follow a glittering career in oil, or gold, or academia, but in keeping strongly alive my interest in the earth, for all the decades that have passed since he taught me. It is with the deepest gratitude for his wisdom, kindness and friendship, that I dedicate this book to him—the longest of all my essays, and thirty-five years late, but well meant all the same.

Index

Page numbers in *italics* refer to illustrations.

Copyright and Illustration Acknowledgments

Table of Strata, William Smith Medal, and Wollaston Medal: With thanks to the Geological Society of London.

Portion of first page of Genesis: With thanks to Dr. John Fuller.

Late Jurassic paleogeographic map by C. R. Scotese, PALEOMAP Project (www.scotese.com). For further information, please consult C. R. Scotese, *Atlas of Earth History*, Vol. 1, *Paleogeography*, PALEOMAP Project (Arlington, Tex.: 2001), p.52.

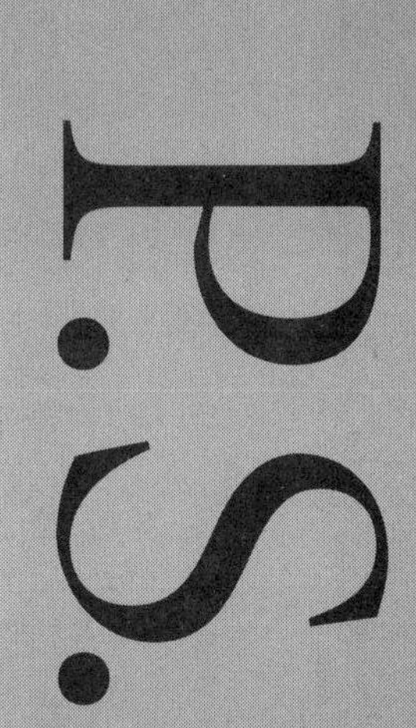

About the author

About the book

Read on

Insights,
Interviews
& More . . .

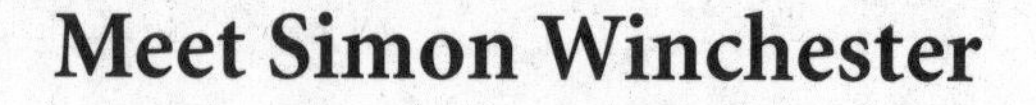

Meet Simon Winchester

AUTHOR, JOURNALIST, AND BROADCASTER Simon Winchester has worked as a foreign correspondent for most of his career. Before joining his first newspaper in 1967, however, he graduated from Oxford with a degree in geology and spent a year working as a geologist in the Ruwenzori Mountains in western Uganda and on oil rigs in the North Sea.

His journalistic work, mainly for *The Guardian* and *The Sunday Times*, has seen him based in Belfast; Washington, D.C.; New Delhi; New York; London; and Hong Kong, where he covered such stories as the Ulster crisis, the creation of Bangladesh, the fall of President Marcos, the Watergate affair, the Jonestown Massacre, the assassination of Egypt's President Sadat, the death and cremation of Pol Pot, and the 1982 Falklands War. During the Falklands conflict, he was arrested and spent three months in prison in Ushuaia, Tierra del Fuego, on spying charges. Winchester has been a freelance writer since 1987.

Setsuko Winchester

He now works principally as an author, though he contributes to a number of American and British magazines and journals, including *Harper's*, *Smithsonian*, *National Geographic*, *The Spectator*, *Granta*, the *New York Times*, and *The Atlantic*. He was appointed Asia-Pacific editor of *Condé Nast Traveler* at its inception in 1987, and later became editor-at-large. His writing has won him several awards, including British Journalist of the Year.

He writes and presents television films on a variety of

historical topics—including a series on the final years of colonial Hong Kong—and is a frequent contributor to the BBC radio program *From Our Own Correspondent.* Winchester also lectures widely—most recently before London's Royal Geographical Society (of which he is a Fellow) and to audiences aboard the cruise liners *QE2* and *Seabourn Pride.*

His books cover a wide range of subjects: the remnants of the British Empire, the colonial architecture of India, aristocracy, the American Midwest, his months in an Argentine prison on spying charges, his description of a six-month walk through the Korean Peninsula, and the Pacific Ocean and the future of China. More recently he has written *The River at the Center of the World,* about China's Yangtze River; the bestselling *The Professor and the Madman,* which is to be made into a major motion picture by distinguished French director Luc Besson; *The Fracture Zone: My Return to the Balkans,* which recounts his journey from Austria to Turkey during the 1999 Kosovo crisis; and the bestselling *The Map That Changed the World,* about the nineteenth-century geologist William Smith. His recent books *Krakatoa: The Day the World Exploded: August 27, 1883* (April 2003) and *A Crack in the Edge of the World: America and the Great California Earthquake of 1906* (October 2005) both have been *New York Times* bestsellers and have appeared on numerous best of and notable lists.

Simon Winchester lives in New York City and has a small farm in the Berkshires in Massachusetts. Mr. Winchester was made an Officer of the Order of the British Empire (OBE) by Her Majesty The Queen in 2006. He received the honor in a ceremony at Buckingham Palace.

“Mr. Winchester was made an Officer of the Order of the British Empire (OBE) by Her Majesty The Queen in 2006.”

William Smith's New Legacy

WHAT FOLLOWS is a testament to the frequently unacknowledged power of readers—in particular to the readers of this book and to whom, by way of this brief epilogue, I offer my most sincere thanks. For it is now clear that some of those who became caught up in the story of William Smith were behind the construction in England of two very tangible and very different memorials to him—two buildings in two separate corners of the British Isles that, one imagines, will now endure for a very long time.

In both places, and as a direct result of these readers' efforts, budgets were written, blueprints were drawn, contractors were hired, stockpiles of oak and pine and sheetrock and stone were assembled, sledgehammers and fretsaws and screwdrivers were employed, until eventually, and with a good deal of pomp in one place and more than a little circumstance in the other, two opening ceremonies were performed—and all because what you hold in your hands prompted a small number of people to pick up their telephones and exclaim to friends with imagination and influence: *Wouldn't this be a good idea?*

> "Some of those who became caught up in the story of William Smith were ultimately behind the construction in England of two very tangible and very different memorials to him. . . ."

The first of these two events, prompted by the first of these telephoned questions, took place in London, at the immense marble palace on Piccadilly where William Smith's first map had been hung since 1874.

The Geological Society of London—a body that in the early nineteenth century consistently denied Smith membership, in large part because of his lowly social status—had nonetheless after his death placed his map foursquare upon the walls of their Burlington House headquarters. The rationale was that, however inferior his standing might have been in the salons of London's haut monde, Smith's intellectual accomplishments were unrivalled and, moreover, he had become the deserving first recipient of what would become the greatest honor of the Society itself, the Wollaston Medal. It would have been churlish, to say the least, not to put his map on display.

However, an all-too-familiar fate awaited it. The map, eight feet by six and surrounded by a wooden frame of dour Victorian solidity, was hung very poorly—suffering the ignominy of many an unwanted painting given to a country squire with too many hanging-places from which to choose. It was never to be placed in the Burlington House lumber-room, true, nor in one of the bathrooms (which are famous for their Silurian and Mesozoic marbles), but it was placed inconveniently high up on a staircase on which very few ever lingered, and it was shrouded, supposedly for protection against the sunshine, by thick blue velvet curtains. As a consequence, hardly anyone ever saw it—a melancholy situation that became much more so after this book was published and visitors to London, usually from the United States, began to turn up unannounced at the Geological Society offices and ask: *Where was William Smith's map?*

A few of these visitors were fortunate and were admitted and shown upstairs, whereupon a Society official appeared and drew back the drapes with a flourish, to general amazement and delight. Most, ▶

"The map, eight feet by six and surrounded by a wooden frame of dour Victorian solidity, was hung very poorly—suffering the ignominy of many an unwanted painting given to a country squire with too many hanging-places from which to choose."

William Smith's New Legacy ***(continued)***

however, went away disappointed. The entrance to the Society in those days was a side door beneath an archway. Supplicants had to press a button on an electric box and explain, to an often harried assistant whose inaudible replies were squawked out across the Piccadilly traffic, just why they wished to gain entrance to what was, after all, a private Society, and why it was necessary to disturb the routines of those working within. Americans in particular have a robustly democratic approach to the great icons of history, which they see as being the property of all humankind; some at the Society, in the immediate aftermath of the publication of this book, made it abundantly clear that they thought otherwise.

However, in the run-up to the bicentenary of the Geological Society, in 2007, the atmosphere changed, and markedly. A new president assumed power: Richard Fortey, an author of great distinction himself and a geologist who felt unreservedly proud of his calling. He let it be known that he wanted all who wished to glory in geology's treasures and delights—particularly those treasures held in the archives or on the walls of what was the oldest professional geological body in the world—to be able to do so.

> "In the run-up to the bicentenary of the Geological Society, in 2007, the atmosphere [toward Smith] changed, and markedly."

Accordingly, and after listening to the pleas of readers of this book who wanted to see Smith's map in particular, a plan was hatched. First, one should no longer be compelled to scuttle into the Society through the side door, but instead be allowed to enter through the long-unused great front door on Piccadilly itself. Second, there should be no further bother with an electrical squawk-box: one should simply open the front door and come in. And third, and most important for this story, anyone who came in should now be able to see, resplendent in a brand new frame and in plain view at the very bottom of the great east staircase, William Smith's extraordinary map. Moreover, and in a demonstration of the conciliatory nature of time, the *other* map, the plagiarized version that was to be the cause of Smith's long fall into penury and oblivion, should be rehung, too, and placed on the wall beside it.

So the necessary funds were eventually found—mainly thanks to a benefactor for the map itself, a wealthy cartographer who singlehandedly paid for it to be restored and rehung. In February 2007, the new entranceway, with slabs of all the prettiest rock types found in the British Isles on prominent display, was formally dedicated. All London was delighted—especially at being able to see these impeccable examples of what are inarguably commercial history's most significant surviving geological maps.

And not a single visitor has been turned away since: William Smith's map is now one of the most popular way stations on any discerning tourist's progress around the British capital, and much satisfaction is expressed that one of the icons of human knowledge and development has now been, in the most democratic sense, given its due place in and to the world. And all thanks to the press of readers' interest and to the flexible mind of a writer-president well aware of the singular power of readers' minds.

"William Smith's map is now one of the most popular way stations on any discerning tourist's progress around the British capital."

The other construction is due not to the press of many, but to the enthusiasm of ▸

one. He is a member of the British aristocracy, a geology enthusiast with the unforgettably impressive name of Robin Evelyn Leo Vanden-Bempde-Johnstone, properly known as the Fifth Baron Derwent.

Lord Derwent, who lives in Hackness Hall, a magnificent Georgian pile in Yorkshire, is a direct lineal descendant of the man whom most believe rediscovered William Smith after his long period of oblivion. In the late 1820s, Sir John Johnstone employed Smith as his estate surveyor and had him create a colored geological map of the policies that surround the great hall. That map survives still, hanging in the gunroom. The present Lord Derwent, a businessman, came to recognize its full significance only when he read this book—and when he came to know the story of how his own ancestor had played so central a role in restoring Smith to the standing he deserved.

Six miles from Hackness, in the seaside town of Scarborough, was one other tangible relic of Smith's restoration. Standing forlorn and unused beside the town's main railway bridge and overlooking the North Sea (the German Sea, according to Smith's 1815 map) was the curious cylinder-shaped museum known as The Rotunda, which the Johnstone family had financed nearly two centuries before to enable Smith to display his enormous collection of fossils. That the building had survived at all struck Lord Derwent as miraculous; that it had fallen into desuetude was, he also came to realize, something that could be reversed.

For six years, working with the full cooperation of the local town government,

Derwent led the effort to raise the necessary millions to bring the building back into fully functioning condition. Being a well connected member of the House of Lords helped: he was able to persuade no less than Prince Charles, the Prince of Wales, to lay the foundation stone, and in May 2008, the William Smith Museum of British Geology was formally opened—helping at a stroke to make the town of Scarborough, a once-popular bathing resort that stands on one of the most geologically rich sections of the British coastline, into a newly revivified destination for travelers and tourists.

And all because Lord Derwent read a book: this book. Such, one might say, is the power and influence of the written word—and more especially, of course, when the written word becomes what it is designed to be: the *read* word. Read by either the hundreds, or by the single unanticipated individual—all of whom have the power to achieve, as these two instances show, the completely unexpected.

Have You Read? More by Simon Winchester

THE MAN WHO LOVED CHINA

In sumptuous and illuminating detail, Simon Winchester brings to life the extraordinary story of Joseph Needham, the brilliant Cambridge scientist who unlocked the most closely held secrets of China, long the world's most technologically advanced country.

No cloistered don, this tall, married Englishman was a freethinking intellectual who practiced nudism and was devoted to a quirky brand of folk dancing. In 1937, while working as a biochemist at Cambridge University, he instantly fell in love with a visiting Chinese student, with whom he began a lifelong affair.

He soon became fascinated with China, and his mistress swiftly persuaded the ever-enthusiastic Needham to travel to her home country, where he embarked on a series of extraordinary expeditions to the farthest frontiers of this ancient empire. He searched everywhere for evidence to bolster his conviction that the Chinese were responsible for hundreds of mankind's most familiar innovations—including printing, the compass, explosives, suspension bridges, even toilet paper—often centuries before the rest of the world. His thrilling and dangerous journeys, vividly recreated by Winchester, took him across war-torn China to far-flung outposts, consolidating his deep admiration for the Chinese people.

After the war, Needham was determined to tell the world what he had discovered, and he began writing his majestic *Science and Civilisation in China*, describing the country's long and astonishing history of invention and technology. By the time he died, he had produced, essentially single-handedly, seventeen immense volumes, marking him as the greatest one-man encyclopedist ever.

Both epic and intimate, *The Man Who Loved China* tells the sweeping story of China through Needham's remarkable life. Here is an unforgettable tale of what makes men, nations, and, indeed, mankind itself great—related by one of the world's inimitable storytellers.

"A masterful biography. . . . Winchester deftly captures [Needham's] complex personality, a romantic adventurer propelled by intellectual curiosity. . . . Winchester has brought Needham vividly to life." —*Boston Globe*

A CRACK IN THE EDGE OF THE WORLD

Unleashed by ancient geologic forces, a magnitude 8.25 earthquake rocked San Francisco in the early hours of April 18, 1906. Less than a minute later, the city lay in ruins. Bestselling author Simon Winchester brings his inimitable storytelling abilities to this extraordinary event, exploring the legendary earthquake and fires that spread horror across San Francisco and northern California in 1906 as well as the startling impact on American history and, just as important, what science recently has revealed about the fascinating subterranean processes that produced the earthquake—and almost certainly will cause it to strike again.

"In this brawny page-turner, bestselling writer Winchester has crafted a

Have You Read? ***(continued)***

magnificent testament to the power of planet Earth and the efforts of humankind to understand her."
—*Publishers Weekly* (starred review)

KRAKATOA: THE DAY THE WORLD EXPLODED: AUGUST 27, 1883

Simon Winchester details the legendary annihilation of the volcano Krakatoa in 1883 and its lasting world-changing effects, including the creation of an immense tsunami, the release of dust that swirled around the world for years, and the triggering of a wave of murderous anti-Western militancy by fundamentalist Muslims in Java.

"One of the best books ever written about the history and significance of a natural disaster." —*New York Times Book Review*

"A real-life story bigger than any Hollywood blockbuster." —*Entertainment Weekly*

THE PROFESSOR AND THE MADMAN: A TALE OF MURDER, INSANITY, AND THE MAKING OF THE *OXFORD ENGLISH DICTIONARY*

The best-selling tale of madness, genius, and the incredible obsessions of two remarkable men that led to the making of the *Oxford English Dictionary*—one of the most ambitious projects ever undertaken. As word definitions were collected, the overseeing committee led by Professor James Murray discovered that one man, Dr. W. C. Minor, had submitted more than ten thousand. When the committee insisted on honoring him, a shocking truth came to light: Dr. Minor, an American Civil

War veteran, was also an inmate at an asylum for the criminally insane.

"An extraordinary tale and Simon Winchester could not have told it better. . . . A splendid book." —*The Economist*

"The linguistic detective story of the decade."
—William Safire, *New York Times Magazine*

OUTPOSTS: JOURNEYS TO THE SURVIVING RELICS OF THE BRITISH EMPIRE

Simon Winchester, struck by a sudden need to discover exactly what was left of the British Empire, set out across the globe to visit the far-flung islands that are all that remain of what once made Britain great. He traveled thousands of miles to capture a last glint of imperial glory.

"A brilliant and delightful addition to the long and distinguished shelf of British literary odysseys." —Christopher Buckley, *Washington Post Book World*

"Winchester traveled one hundred thousand miles back and forth from Antarctica to the Caribbean, from the Mediterranean to the Far East, and has come up with a fascinating and important book."
—*The Times* (London)

Have You Read? *(continued)*

THE FRACTURE ZONE: MY RETURN TO THE BALKANS

A true portrait of one of the world's most chaotic and beautiful regions that explains why violence has always occurred there—and why it may continue to occur there for years to come.

"A vivid, informative history of the Balkans."
—*Chicago Tribune*

"Scholarly and moving . . . combines historical significance with dramatic insight." —*The Independent*

KOREA: A WALK THROUGH THE LAND OF MIRACLES

Fascinating for its vivid presentation of historical and geographic detail, *Korea* is that rare book that actually defines a land and its people, while providing Winchester's gift for writing about engaging characters in true, compelling stories.

"Immensely readable. . . . Winchester made his journey of over three hundred miles on foot, a remarkable achievement in itself and one that afforded him a unique opportunity to experience both the country and its people at a grassroots level." —*The Guardian*